# The Road

# PRIVATISATION

## From the Colin Garratt Collection

### Described by Colin Scott-Morton

The Transport Treasury

The Transport Treasury

*Reviving the memories of yesterday…*



ISBN 978-1-915281-12-8

First published in 2024 by Transport Treasury Publishing Ltd., 16 Highworth Close, High Wycombe, HP13 7PJ

www.ttpublishing.co.uk

Printed



***Front cover main:*** **The Network SouthEast Thames and Chiltern division was split during the privatisation process to form the separate Thames Trains and Chiltern Railways franchises. One of Chiltern's aspirations was to restore a second main line between London and Birmingham, operating from its Marylebone terminus. This ambition was achieved through infrastructure improvements, increasing both capacity and line speed, and the introduction of Class 168 'Clubman' diesel multiple-units, with main-line standard interiors and 100 mph capability. One of these, unit 168 112, is seen here starting away from Leamington Spa with a London-bound service.**

***Front bottom left:*** **Class 91 locomotives 91 006 and 91 020 stand at the platform ends at King's Cross on 5th March 1997. The 91s - and their accompanying Mark 4 coaches - were introduced by BR from 1989, specifically to work services over the East Coast main line. At privatisation, the InterCity East Coast franchise was awarded to Sea Containers Ltd, trading under the Great North Eastern Railway (GNER) name. The operator decked out its rolling stock in a new livery of dark blue, with a broad red stripe at waist level, and these two machines look particularly smart in these colours.**

***Front bottom right:*** **From 1960, electric services in the Glasgow suburban area were operated by a fleet of 91 Class 303 'Blue Train' electric multiple-units (so-called because of their original Caledonian Blue livery). By the time that Scotrail services were privatised in 1997, numbers had dwindled to 40, as newer stock had gradually displaced them. Their withdrawal was finally completed following the introduction of Class 334 'Coradia' units, built by Alstom between 1999 and 2002. One of these, unit 334 010, sets off from Glasgow Central with a sister unit in tow, both wearing Strathclyde PTE's carmine and cream colours.**

***Rear cover upper:*** **The Central Trains franchise, commencing in March 1997, operated services from Birmingham and Nottingham to Cardiff, Aberystwyth, Shrewsbury, Liverpool, Sheffield, Cambridge and Norwich, as well as suburban services in the West Midlands PTE area. Longer distance workings were initially in the hands of Class 158 diesel units, all of which started out wearing BR Regional Railways livery. From 1999, they began to appear in a striking, predominantly green colour scheme, seen here on an unidentified unit at Nottingham. This early version carries the operator's phone number on the lower bodyside, replaced on later iterations by their website address.**

***Rear cover lower:*** **The 'Pacer' units introduced by BR during the 1980s undoubtedly had 'Marmite' qualities - people either loved them or hated them! With their modified bus bodies mounted on a four-wheel chassis, they could be noisy and rough-riding, but they provided a useful low-cost rolling stock solution for lightly-used rural routes. They also stood the test of time, with the last of them remaining in service until 2020. This view, at Carlisle, shows Class 142 unit 142 026 waiting to depart with a Northern Spirit service to Middlesbrough. Even the most hardened Pacer fan (if such people existed) would probably have found this three-hour journey an ordeal, though!**

***Frontispiece:*** **Following their initial deployment on the South Wales, West of England, and East Coast main lines, BR introduced HST sets on Cross Country routes from 1982. Their operation continued into the privatisation era, although heavily supplemented by newer 'Voyager' units. On the leg between Birmingham and the South West, an InterCity liveried HST starts the descent of the Lickey Incline towards Bromsgrove, with the Malvern Hills visible in the background. This two-mile stretch, with an average gradient of 1 in 37.7, is the steepest sustained main-line railway incline on the British railway network.**

**CONTENTS** **PAGE**

# INTRODUCTION

From nationalisation in 1948 until the 1980s, the British railway system was organised on geographical lines, divided up into regions which largely followed the areas covered by the former 'Big Four' companies. Those regions were - Southern, Western, London Midland, Eastern and Scottish. There was originally also a North Eastern Region, but this was amalgamated with the Eastern in 1967. Each region was responsible for most of what went on within its 'territory' - passenger and freight services, stations, depots, signalling, engineering and infrastructure maintenance - although construction and maintenance of rolling stock was often handled by the major works spread around the country.

The national system was perennially loss-making and supported by taxpayer subsidy, although it became clear over time that some parts of it performed better financially than others. Nevertheless, various initiatives were embarked upon to stem the losses, perhaps the best remembered being the famous - or possibly infamous - Beeching Reshaping report of the 1960s, which resulted in the closure of around a third of the network's total route mileage. The plan was undoubtedly successful in reducing costs, and, arguably, some of the lines closed should probably never have opened in the first place. However, it was less effective in increasing revenue, particularly in the face of growing competition from increased car ownership and domestic air services, so the railways' overall financial position was not greatly improved. To some extent, it also failed to consider adequately the contribution that loss-making routes made to the wider network. Once a prospective traveller was faced with driving several miles to reach a railway station rather than boarding a train locally, it was often a more attractive option to stay in the car all the way to the final destination.

A rather more enlightened - and successful - approach was demonstrated by the sectorisation programme of the early 1980s. This saw a move away from the regional structure, replacing it with one based on 'business sectors', each focusing on a particular market or traffic flow. The sectors were:

- InterCity - covering the major trunk routes, especially the East and West Coast main lines between London and Scotland, the Midland main line, the routes to South Wales and the West Country, and what had been the 'inter-regional' cross country services;

- Network SouthEast (initially the London and South East sector) - responsible for the vast complex of commuter routes into London, from

**The new order on the West Coast main line. Virgin Trains, who took on the InterCity West Coast franchise from March 1997, placed an order for 53 Class 390 'Pendolino' tilting train sets. These were built at Alstom's Washwood Heath Works in Birmingham between 2001 and 2004, in 9-coach formation, with some later extended by two coaches. Their tilting capability allowed sustained 125 mph running on the sinuous West Coast route, bringing some journey time reductions without compromising passenger comfort. This view was taken inside Washwood Heath at a Pendolino launch event, with a brand new set surrounded by the manufacturer's staff. In the left background can be seen a Class 458 unit for South West Trains.**

such far-flung locations as King's Lynn in the east, and Exeter in the west;

- Regional Railways (initially Provincial sector) - operating virtually all the remaining passenger services not covered by the other two sectors;

- Railfreight - little explanation necessary!

The structure was refined over time, with various sub-sectors created, particularly within Railfreight, which had a number of clearly defined markets such as petroleum, coal, etc.

Unique among Europe's railways, the InterCity sector generated an operating profit in 1988, and continued to do so for the next six years. The quality of the product offered to passengers was consistent, with all services formed of 100 or 125 mph air-conditioned stock, and catering - ranging from a buffet up to full restaurant service - provided on all trains.

While Network SouthEast did not achieve the same financial successes as InterCity, it did transform the commuting experience for millions of passengers throughout the south east. Its tenure saw the total modernisation of the Chiltern lines out of Marylebone, electrification of no less than 17 routes, and implementation of the 'Thameslink' cross-London service. New rolling stock was introduced, some in connection with the schemes mentioned above, while other classes replaced older, life-expired equipment. A widespread programme of station improvements was carried out, with the familiar red lamp posts and seats, and new signage, brightening up what had previously been drab and run-down facilities.

Regional Railways also achieved much, with the wholesale replacement of first-generation diesel multiple-units by new 'Pacer' and 'Sprinter' units. The superior performance of the new stock, together with greater reliability and availability, enabled the sector to reduce some journey times and increase frequency on many rural routes, with a consequent increase in passenger numbers.

Few might have imagined, while all this progress was taking place during the 1980s, that the following decade would see even greater and - arguably more disruptive - change, in the form of privatisation of the railway network. The process was complex, involving the fragmentation of the existing passenger sectors into 26 Train Operating Units, ready to be let to private sector franchisees on terms typically lasting seven years. Over the second half of the decade and into the new century, the existing identities gradually disappeared, to be replaced by a dizzying array of new names and liveries, some relatively short-lived themselves as franchises changed hands and businesses restructured. The bulk of the network's infrastructure, including the main stations in major towns and cities, was transferred to a new company, Railtrack, while the Railfreight sector's activities were sold off to a number of private operators. British Rail's vast fleet of passenger rolling stock was distributed among three leasing companies, for onward hire to franchisees.

Whether privatisation was a good or bad thing has been hotly debated over the last three decades, and will probably continue to be so. It is not the purpose of this book, as a picture album, to advance that debate one way or the other, but rather to illustrate the changing face of our railway system during that period of transition. The images that follow are divided into sections, each covering a British Rail sector and the various franchised operations that succeeded it. For the most part, this has been restricted to the first private operator to take over, although there are exceptions. As mentioned above, some of the franchises changed hands fairly quickly; most have done so in the intervening years, and almost thirty years on, the structure has been largely dismantled, replaced by a system of management contracts under which private companies are paid a fee to operate services, with the government keeping the fare revenue and effectively taking the financial risk. How we got from those first franchises to the current position is a further story in itself, but outside the scope of this book.

Finally, I would like to express my gratitude to everyone at Transport Treasury for their help and efforts in sourcing and scanning the images that follow.

**Colin Scott-Morton**

BR InterCity sector

Top left. In addition to the trunk routes between London and major cities, InterCity also operated an extensive Cross Country network radiating from Birmingham to the south coast, the West Country, South Wales and Scotland. With large sections of these routes not electrified, services were worked by a mixture of HST sets and locomotive-hauled formations. One of the latter is seen here, passing Aynho Junction south of Banbury, with a working bound either for Brighton or Poole. Motive power was provided by an unidentified Class 47 locomotive - with a particularly smoky exhaust - while passenger accommodation comprised one first class coach, five open seconds, and a brake second at the rear, all air-conditioned Mark 2 vehicles. The first class coach behind the locomotive included a small buffet area at its inner end.

Bottom left. Another Class 47 hauled Cross Country working is seen here, heading south from Manchester Piccadilly station. The train formation is similar to that in the previous picture except that, unusually, it appears to include two brake second vehicles and only four open seconds. The locomotives rostered for these services were usually of sub-class 47/4, with extended range fuel tanks, allocated to Bristol Bath Road depot.

Above. On the electrified West Coast main line, InterCity services were operated by Class 86, 87 and 90 locomotives, usually in 'push-pull' mode with a Driving Van Trailer (DVT) marshalled at the London end of the train. In this view, a Class 86 heads north leading eight Mark 2F coaches, a Mark 3 catering vehicle, and a DVT at the rear. This formation suggests that it is one of the 'shorter range' services between Euston and Birmingham/Wolverhampton. Most longer distance workings, to Liverpool, Manchester and Glasgow, were normally full Mark 3 rakes powered by Class 87 or 90 locomotives, allowing 110 mph running rather than the 100 mph to which Mark 2F vehicles were restricted.

INTERCITY

Left. InterCity's electric fleet development spanned more than two decades, with Class 86 introduced from 1965, Class 87 from 1973, and Class 90 from 1987. There is a direct lineage running through them, however, with each being a derivative of the earlier class, featuring newer technology and uprated performance. The Class 90s were built by British Rail Engineering Ltd (BREL) at Crewe Works between 1987 and 1990, and had a power output of 5,000 hp and a maximum speed of 110 mph. Although a mainstay of West Coast main line operations, they were not irregular performers on services from King's Cross, particularly those to Leeds, following electrification of that route in 1988. One such working is seen here, heading south at Sandal and Agbrigg, between Wakefield and Doncaster.

Above. For the full East Coast electrification scheme, from King's Cross to Leeds and Edinburgh, a fleet of Class 91 locomotives and Mark 4 coaches were provided, the former built by BREL and the latter by Metro-Cammell. Both were designed for 140 mph operation, and it was originally planned to market the service as 'InterCity 225' (225 km/h being the metric equivalent of 140 mph). However, concerns over signal sighting at this speed led to a decision that it would only be permitted with in-cab signalling, with 125 mph remaining the maximum with conventional lineside signals. The Class 91 was of unconventional design, having a streamlined cab at one end, and a flat-fronted one at the other. This allowed it to 'blend' visually with the Mark 4 coaches in normal passenger service, as seen in this view of a northbound working at Essendine. On the rare occasions that the locomotives operated with the flat, number 2, end leading, maximum speed was restricted to 110 mph.

INTERCITY

Top left. Originally developed in the early 1970s, the 125 mph diesel-powered High Speed Train (HST) remained a mainstay of InterCity's non-electrified routes into the 1990s. These included the former Western Region routes to the west of England and South Wales, the Midland main line, the Cross Country network, and the lines north of Edinburgh to Inverness and Aberdeen. On the latter route, power car 43 116 leads a southbound HST past Kerr's Miniature Railway in Arbroath. This 10¼ inch gauge railway, opened by Matthew Kerr in 1935, ran alongside the main line for half a mile in West Links Park. It was the oldest miniature railway in Scotland, and had been a popular, family-run attraction in the town for many years. Sadly, falling passenger numbers led to its permanent closure in late 2020.

Bottom left. On-train catering has always been an important element of long-distance services, albeit one that is generally loss-making when considered in isolation, and which has had a 'mixed' reputation for quality over the years. As part of the sectorisation process in the 1980s, a dedicated unit - InterCity On Board Services (ICOBS) - was set up to handle catering on InterCity trains. ICOBS embraced new technology and introduced many innovations, including 'modular' handling of stores, leading to greater efficiency and flexibility. As a result, InterCity soon achieved a well-deserved reputation for the high quality of its catering, defusing once and for all the old 'curly sandwich' jokes. One of its most popular offerings was the 'Great British Breakfast', which is seen here being served to passengers in a Mark 4 vehicle.

Above. The InterCity East Coast franchise - covering the routes from King's Cross to Leeds, Edinburgh, Glasgow, Inverness and Aberdeen - was awarded to Sea Containers Ltd in March 1996, with operations starting under the Great North Eastern Railway (GNER) name on 28th April. The company took on the former BR fleet of Class 91 locomotives, Mark 4 coaches and HST sets under leasing agreements, and decked these out in a new livery of dark blue with a broad red stripe at waist level. This view, at King's Cross on 5th March 1997, shows Class 91 locomotives 91 006 and 91 020 standing at the platform ends in the new colours. As can be seen, there was some variation in the application of the livery in the early days, with one machine carrying the GNER logo in gold, and the other in white. The gold version was ultimately settled on as standard.

Above. HSTs had been associated with the East Coast main line since 1978, when they first began to displace the Class 55 'Deltic' locomotives from front-line services. They remained in service from privatisation in 1997, and continued to ply the route with GNER and its successors for another 22 years, until finally replaced by Hitachi-built Class 800 'Azuma' sets. In GNER days, an HST set awaits passengers at King's Cross, prior to another trip north.

Top right. At the opposite end to the Class 91 locomotive on an 'InterCity 225' formation was the Driving Van Trailer (DVT). This essentially comprised a Mark 4 coach bodyshell - containing parcels and luggage space rather than passenger accommodation - with a Class 91-style cab front at the outer end. They were normally marshalled at the London end of the train, which would be driven from this vehicle on southbound workings. Thirty-two DVTs were built by Metro-Cammell for the East Coast electrification, with manufacture of the bodyshells subcontracted to the Italian company Breda. One of the vehicles is seen bringing up the rear of a northbound working as it passes the flat crossing north of Newark. This carries the Nottingham to Lincoln line across the East Coast main line, and is the last remaining location where two standard gauge routes cross on the level on the UK national network.

Bottom right. York station - one of the major 'hubs' on the GNER network - is the setting for HST power car 43 039. This is another example of a vehicle displaying the non-standard and short-lived GNER logo in white. The door at the right-hand end of the vehicle provided access to the original guard's accommodation when the HST sets were first introduced. However, engine noise was found to be excessive in these areas, and so from 1980 alternative accommodation was provided in newly built Trailer Guard Second (TGS) vehicles, effectively a Trailer Second with the end vestibule and one seating bay replaced by a guard's compartment. These vehicles were always marshalled next to the power car - which retained luggage accommodation and brake controls.

100

GNER
43039

40
40

Top left. When the Channel Tunnel services were being planned, it was proposed to operate 'Regional Eurostar' trains between Glasgow, Manchester, Paris and Brussels. A fleet of seven Class 373/3 'North of London' 14-vehicle Eurostar sets was built to operate these trains, the 'Three Capitals' sets working London services having 18 vehicles. For a number of reasons, including the privatisation of British Rail and lower than anticipated passenger numbers on the Three Capitals workings, Regional Eurostar was not considered viable and the services never ran. Between 2000 and 2005, GNER used some of the North of London sets to provide extra capacity on services to York and Leeds - branded 'White Rose' - and these had the company's livery applied by the use of vinyl wraps. On 20th December 2001, sets 3303/3304 approach Doncaster with a northbound working.

Bottom left. The number 1, 'streamlined end', driving cab of a Class 91 locomotive, showing the driver's seat and controls. The handle in a quadrant to the left of the seat operates the brake, while the power controller is on the centre console to the right, just above the 'GEC' logo. The slightly raised 'pedal' in the floor is the driver's safety device. The main gauges are - from left to right - brake cylinder pressure (top), main reservoir pressure (bottom), brake pipe pressure, speed selector, and speedometer. Visible through the windscreen is the number 2, 'blunt end', of a sister locomotive.

Above. Electric services commenced over the Great Eastern main line between London and Norwich in May 1987, using a fleet of Class 86 locomotives and predominantly Mark 2 coaches with Mark 3 catering vehicles. This route formed the core of the InterCity Anglia franchise, which was awarded to GB Railways Group PLC, trading as Anglia Railways, in January 1997. Anglia also operated services to Harwich, together with some regional services throughout an area bounded by Felixstowe, Lowestoft, Great Yarmouth, Sheringham, Ely, Peterborough and Cambridge. A turquoise livery with white waist level stripe and white logos was applied to the operator's rolling stock, as seen on this London-bound service hauled by locomotive 86 218. The Mark 3 catering car is the third vehicle in the formation.

Above. London to Norwich services operated in 'push-pull' mode, with the Class 86 locomotive marshalled at the London end, and a Driving Brake Standard Open (DBSO) vehicle at the other. The DBSOs were converted from Mark 2F Brake Standard Open coaches, by the addition of a driver's cab at the non-passenger end. The first ten conversions were carried out in 1979, with the vehicles deployed on Class 47 worked InterCity services between Edinburgh and Glasgow, and a further four were added in 1985/6. In 1990, following the introduction of Class 158 diesel multiple-units in Scotland, the DBSOs were transferred to Norwich Crown Point depot for use on the Great Eastern route. Some modifications were made to the vehicles at this time, including removal of the end gangways - which had been retained during the original conversion work - and the provision of a full width driving cab. In this view between Shenfield and Colchester, a London to Norwich service storms along the Great Eastern main line behind DBSO 9709.

Top right. In order to provide some direct services from London to Sheringham, Great Yarmouth and Lowestoft - beyond the limit of electrification at Norwich - Anglia acquired eight Class 170 'Turbostar' three-coach diesel units in 1999/2000. The Class 170s were built between 1998 and 2005 by Adtranz and their successor Bombardier Transportation, at Litchurch Lane Works in Derby. The fleet totalled 139 units, in both two- and three-coach formations, and the sets found employment with many operators around the UK. They had a maximum speed of 100 mph, enabling them to maintain the same timings as the electrically worked services on the Great Eastern main line, and their comfortable interiors, with seating in a two-plus-two layout, was well suited to longer distance runs. Anglia's units initially had a small buffet counter in the centre coach as well, although this was later removed to provide additional seating. Externally, they wore a variation of the original livery, with 'toned down' turquoise covering just the lower bodyside, as seen on unit 170 206 standing at Liverpool Street station.

Bottom right. On 30th July 1984, a push-pull service from Edinburgh to Glasgow, led by a DBSO, was derailed near Polmont after striking a cow which had strayed on to the line through a damaged fence. The train was travelling at around 85 mph just prior to the collision, and all six coaches and the Class 47 locomotive at the rear were derailed, with the leading vehicles sustaining severe damage. Thirteen passengers were killed, and a further seventeen people, including the driver, were seriously injured. One of the factors considered by the inquiry was whether the fact that the leading vehicle was a passenger coach, rather than a heavier locomotive, made the train more susceptible to derailment in these circumstances. Having examined all of the evidence, the inspecting officer concluded that there was nothing inherently unsafe about push-pull operation, but recommended that obstacle deflectors should be fitted to the leading bogies on DBSO vehicles. This feature can be clearly seen on Anglia DBSO 9709, standing at Ipswich at the rear of a London-bound service.

170 206

Ipswich
Anglia
9709

# Gatwick Express

**Above. Gatwick Airport has been served by rail since its opening in 1958, initially by portions detached from - or attached to - trains between London and the south coast. From May 1984, a dedicated non-stop service was introduced between London Victoria and the airport, branded 'Gatwick Express', with a 15-minute frequency and 30-minute journey time. In the best traditions of the old Southern Railway and BR Southern Region, the stock provided for this new service was all second-hand! Trains were formed of converted Mark 2F coaches in fixed two- and three-coach sets (allowing for a variety of train lengths), topped and tailed by a Class 73 Electro-diesel locomotive and a Class 489 Gatwick Luggage Van (GLV). These latter vehicles were converted from former multiple-unit driving cars, with the interiors stripped out and three pairs of double loading doors provided on each side. The driver's cab was retained at the outer end, as was the powered bogie underneath. The service became the responsibility of InterCity in April 1985, and it was the first part of that sector to become a 'shadow franchise', ready to be transferred to the private sector, in October 1993. Wearing the new private operator's livery, a service from Victoria is seen here approaching journey's end, with Horley station in the background.**

**Top right. The InterCity Great Western franchise, covering routes from London to the West Country, South Wales, Worcester and Hereford, was awarded to Great Western Holdings, trading as Great Western Trains, on 4th February 1996. The holding company was formed in 1994, specifically to bid for rail franchises during the privatisation process, and its shareholding was split between former BR managers (51 per cent), FirstBus (24.5 per cent) and 3i (24.5 per cent). A new livery of green and ivory was developed for the operator's rolling stock, together with a logo showing a stylised Merlin Falcon, as seen here on an unidentified HST power car.**

**Bottom right. Livery transition at Penzance. HST power car 43 186, *Sir Francis Drake,* looks resplendent in the new Great Western Trains colours, while the Mark 3 coaches behind remain in the final version of the InterCity scheme. Mixed formations such as this were as common during the franchising process as they had been during the changeover from the BR regions to business sectors a decade earlier. This power car originally entered service in 1982, and received its name on 31st July 1987. It was later formed into one of Great Western's short, 2+4, 'Castle' sets, and renamed *Taunton Castle*. It remained in service at the time of writing, but these sets were due for withdrawal during 2024.**

# Great Western

Great Western

Top left. A late afternoon panoramic view across Penzance station, the most westerly outpost of the Great Western network, and the southernmost railway station in Great Britain. In the right background can just be seen the train shed, designed by the original Great Western Railway's engineer, William Lancaster Owen, while behind is the dome of the town's Market House, and, at far left, the tower of St Mary's church. Two of the station's four platforms are occupied by HST sets, that on the right appearing to have recently arrived from London, after a journey of some five hours plus. Both are smartly turned out in the company's 'Merlin' livery. Power car 43 017, at the head of the left hand train, was one of the earliest examples of the class, emerging from Crewe Works in September 1976.

Bottom left. A Great Western HST pauses at Kemble with a Cheltenham to Paddington service. Alone among the private operators, Great Western retained the 'InterCity' branding for a period, and its rolling stock carried both the new name and the original BR style titles at alternate ends of the vehicles, as seen here. Kemble, on the line between Swindon and Cheltenham, was originally a junction, with branch lines to Tetbury to the south, and Cirencester to the north. Both were closed under the Beeching reforms, that to Tetbury completely in April 1964, while Cirencester lost its passenger service at the same time, but remained open for freight until October 1965. A further economy measure saw the line from Kemble to Swindon reduced to single track in 1968 - the convergence is just visible here in front of the tunnel mouth - but this was later reversed with the restoration of double track in 2014.

Below. From June 1964, the name "Golden Hind" was applied to the first train of the day from Penzance to Paddington, and to a balancing down working in the evening. Times have varied over the years, but typically the up service left the Cornish town at around 05.00, with the evening Paddington departure at around 18.00. At its launch, it broke all previous records for the Plymouth to London timing, encouraging one passenger to purchase the first ever first class annual season ticket from Taunton to Paddington! The name survived into the privatisation era, and the London-bound working is seen here passing Newbury in July 1997, led by HST power car 43 008.

Above. 'Motorail' services, which enabled travellers to avoid long road journeys by conveying their cars to their destination by train, first operated in the UK in the 1950s. The network expanded over the years, covering such far-flung locations as Penzance and Inverness. However, by the mid-1990s the comprehensive motorway network had reduced road journey times significantly and this, combined with the ongoing privatisation process, led to the withdrawal of the last services in 1995. Great Western ambitiously reintroduced them four years later as part of the "Night Riviera" sleeper service to Penzance, using eight converted General Utility Vans (GUV), numbered 96602-9, to carry cars. Unfortunately, the service was not a commercial success, and was withdrawn again in 2005. This view at Paddington shows two of the Motorail vans behind a Class 47 locomotive. In 1998, First Group bought out the remaining shares in Great Western Trains, and restyled the operation as 'First Great Western', with a revised green and gold livery applied to rolling stock, as seen here.

Top right. The InterCity West Coast franchise, covering the routes from London to Birmingham, Manchester, Liverpool, Holyhead, Glasgow and Edinburgh, was awarded to a joint venture of Virgin Group and Stagecoach, trading as Virgin Trains. Franchise award took place in January 1997, with operations commencing on 9 March. At its inception, the new operator took on the existing InterCity rolling stock fleet, comprising Class 86, 87 and 90 electric locomotives, Mark 2 and 3 coaches with Driving Van Trailers (DVT), and a number of HST sets, the latter primarily for services to Holyhead. A new livery - primarily red with charcoal at vehicle ends, and white bodyside stripes - was applied to the operator's rolling stock, although as with other franchises it took some time before full coverage was achieved. Class 87 locomotive 87 006, *George Reynolds*, wears the new colours as it stands at Crewe coupled to a Mark 3 coach, still in InterCity livery. George Reynolds, who died in 2001, was a journalist and British Rail public relations officer, remembered particularly for his role representing the railway in Scotland.

Bottom right. In addition to retaining InterCity's rolling stock in its early years, Virgin also carried forward the on-board catering offering. Regular travellers were pleased to see the retention of such delights as the 'Great British Breakfast', seen here being served in a Mark 2 open first vehicle. The striking Virgin colours and brand are prominent everywhere, from the seat moquette, antimacassars and carpet, to the staff uniforms.

Virgin West Coast

**Below. Classic West Coast push-pull formation in Virgin livery - Class 90 locomotive 90 015 propels a southbound service along the West Coast main line, controlled from the DVT at the leading end. These trains were normally formed this way, with the first class vehicles and the DVT marshalled at the London end, allowing the locomotive to be detached and run to Willesden depot for maintenance when required. On closer inspection, the operator's colours are not quite universal on this train. The catering vehicle - the sixth coach from the rear - still wears the InterCity scheme.**

**Top right. The West Coast franchise included a commitment to replace the original electric locomotives and coaching stock, and Virgin fulfilled this with their order to Alstom/Fiat Ferroviaria for 53 Class 390 'Pendolino' sets. These were 9-coach fixed formation electric trains - with some later extended to 11 coaches - built at Alstom's Washwood Heath Works in Birmingham between 2001 and 2004. The Pendolinos have a maximum speed capability of 140 mph, but the retention of lineside signals during the scaled back West Coast route modernisation programme around the turn of the century means that they are restricted to 125 mph in service. A significant feature of the Pendolinos is their ability to tilt, enabling them to negotiate curves at higher speeds without causing discomfort to passengers. Of course, British Rail developed a tilting train - the Advanced Passenger Train (APT) - in the late 1970s, but a lack of funding and political will resulted in the project being cancelled in 1986. It is sad to reflect that, had this not been the case, tilting trains could have been working regularly over the West Coast main line more than a decade earlier than they were. At a launch event at Crewe Heritage Centre on 16th June 2002, Pendolino set 390 006 poses alongside preserved APT set 370 003.**

**Bottom right. The formation of the 9-coach Pendolino sets comprised three first class and six standard class vehicles; the 11-coach sets included an additional first and standard class vehicle. Both driving cars are powered, as are four of the intermediate vehicles in the shorter sets, and five in the longer ones. Each powered vehicle carries two Alstom 4 EJA 2852 traction motors rated at 570 hp, giving a total power output of 6,840 hp (9-coach sets) and 7,980 hp (11-coach sets). The first class driving vehicle at one end of the set includes a kitchen, providing full meal service to first class passengers. For standard class travellers, Virgin decided to move away from the traditional buffet car in favour of an on-board shop, selling newspapers and magazines, etc, in addition to food and drink in a self-service arrangement. Hot snacks and drinks were still available over the counter from the steward. The shop was located in the third standard class vehicle from the end of the set, and an example can be seen in this view.**

InterCity
370003
Virgin

Above. A Pendolino shows its tilting capability to advantage as it leans to the curve heading south from Penrith. Each vehicle can tilt up to eight degrees from the horizontal, which equates to a 15-inch difference in height above the track from one side of the coach to the other. In addition, track on the UK railway network is canted by up to six degrees on curves and this, combined with Pendolino's tilt, allows the train to negotiate bends at speeds 20 per cent higher than would otherwise be the case. A new livery was developed for the Pendolinos (and also used on the company's diesel 'Voyager' units), comprising silver bodysides with a black band along the windows, red roofs and cabsides, and a white curved band separating the red and silver areas at the unit ends. The once obligatory warning yellow covered the lower nose end, with a Virgin logo in the centre.

Top right. Between the East and West Coast routes, the Midland Main Line runs north from London's St Pancras, serving St Albans, Luton, Bedford, Kettering and Leicester, before reaching destinations at Nottingham, Derby and Sheffield. Some services continued to Burton-upon-Trent, Matlock, Barnsley, Leeds, York and Scarborough, but unlike its neighbours either side, trains from St Pancras had long ceased to reach Scotland, and the route was considered something of a 'Cinderella' by some. Nevertheless, it was among the earlier franchises to be let, with National Express taking on a 10-year contract from 28th April 1996. A striking new livery was applied to the company's rolling stock, comprising a teal green upper body above a broad beige stripe, with three narrow tangerine bands running along the bottom of the teal section. HST power cars carried the operator's name in bold type, together with the leaping stag logo, as seen here on vehicle 43 047.

# Midland Main Line

Above. The Midland Main Line livery sat particularly well on the HST sets, as seen in this view of one passing Syston with a morning London-bound working. Syston, around 4½ miles north of Leicester, is the junction with the line via Melton Mowbray to Peterborough. The station here - which lies just beyond the overbridge in the background - originally opened in 1840, was closed in 1968, and reopened again in 1994 as part of phase one of the 'Ivanhoe line' project. It has a single platform facing on to the reversible slow line - the right-hand of the three lines here - which runs from just north of Leicester to Syston junction, where it becomes double track.

Left. Sidelit by late afternoon sunshine, another southbound Midland Main Line HST passes between fields of oilseed rape to the north of Market Harborough, with the villages of East Langton and Church Langton visible on the hillside behind. HSTs have been associated with this route since 1982, when falling passenger numbers elsewhere allowed the transfer of some sets from the Western Region and the inter-regional pool. Although none of the line was cleared for 125 mph running, the HSTs represented a better quality product for passengers than the Class 45-hauled rakes of Mark 2 coaches which had served hitherto, as well as enabling some minor journey time improvements. During the InterCity years, the transferred sets were shared between the Midland and East Coast main lines, but in the run-up to privatisation a dedicated 'Midland' pool was created, comprising 31 power cars and 14 sets of coaches. The sets were based at Neville Hill depot, and the extension of some Sheffield services to and from Leeds allowed them to reach this location for regular examinations and maintenance.

Below. In April 1997, Midland Main Line became the first operator to order Class 170 'Turbostar' diesel units from Adtranz, with the first delivered in 1998, and entering service in May 1999. The initial order was for 13 units, later increased to 17, and while all were originally ordered as two-coach sets, the first ten were subsequently made up to three coaches. The number series was 170 101 to 110 (three-coach sets) and 170 111 to 117 (two-coach sets). These units stayed with the company until 2004, when they were transferred to Central Trains - another National Express operated franchise - and replaced on the Midland route by Class 222 'Meridian' sets. Prior to entering service, brand new unit 170 101 heads south near Market Harborough while on test.

Midland Main Line deployed its Class 170 units primarily on stopping services from St Pancras to Leicester, Derby and Nottingham, allowing the HSTs to be more efficiently diagrammed on the longer distance, limited stop workings.  They also worked Saturday services to York, which later operated all year, extended during the summer to Scarborough.  Until 2004, the company provided direct services between London and Matlock, in the Peak District, and these were also in the hands of the 170s.  Unit 170 103 looks smart in the operator's teal and tangerine colours as it threads an unidentified tree-lined cutting in bright sunshine.

# Virgin Cross Country

Above. In addition to the West Coast main line, the Virgin Group was also awarded the InterCity Cross Country franchise in November 1996, with operations over this extensive network commencing on 5th January 1997. The company inherited the BR sub-sector's rolling stock fleet of Class 47 diesel and Class 86 electric locomotives, Mark 2 coaches, HST sets, and Class 158 'Express Sprinter' diesel multiple-units, and initially continued to operate a similar service pattern. On 22nd April 1998, Virgin-liveried Class 47 locomotive 47 807, *The Lion of Vienna*, awaits departure time at York with the 11.43 service to Poole. 'The Lion of Vienna' was the name conferred upon the Bolton Wanderers forward, Nat Lofthouse, after scoring his second goal in England's 3-2 victory over Austria on 25th May 1952. The locomotive was named in his honour in a ceremony at Bolton station on 20th April 1998, just two days before this view was taken.

Below. Although technically a Virgin West Coast service, the workings between Euston and Holyhead were operated by Cross Country HST sets in the early days of the franchise. In this view south of Crewe, a London-bound working from the Anglesey port creates a spectacular exhaust haze from its two power cars as it accelerates away from the scheduled stop. Reflecting different levels of demand on the cross-country routes, the sets allocated by BR to these services had only a single first-class coach, then a catering vehicle (second from the rear in this view) and five standard-class coaches. By contrast, those on London routes had two first-class vehicles and four standard (Great Western) or five (East Coast).

Above. Another Cross Country HST on the London to Holyhead route, this time in the spectacular setting of the 13th century Conwy Castle, on the North Wales coast line. An amusing story - possibly true - relates that an American tourist, on looking at this scene, was heard to remark, "all very nice, but you built the castle much too close to the railroad"! The 'barbican' at the right-hand side of the picture is the entrance to the bridge carrying the line across the River Conwy, designed by Robert Stephenson and opened in 1849. This bridge is of 'tubular' design, with each line carried inside an enclosed, self-supporting box-girder structure, fabricated from wrought iron. A similar structure was employed for the longer Britannia Bridge across the Menai Strait to Anglesey, further west, but a fire in 1970 damaged its tubes beyond repair, and the bridge was rebuilt with truss arches and twin decks carrying both the railway and the A55 road. The Conwy bridge is now the only remaining example of this tubular design, and is Grade 1 listed.

Top right. As with the West Coast franchise, Cross Country was also committed to replacing all of the ex-BR rolling stock during its term. In 1998, it placed an order with Bombardier Transportation in Belgium for 78 'Voyager' diesel electric multiple-units (DEMU). Of these, 34 are designated Class 220, and are four-coach units. The remaining 44 are Class 221 - also known as 'Super Voyagers' - and have a tilting capability similar to the West Coast Pendolinos. Four of the Class 221 sets also have four coaches, while the rest are in five-coach formation. Seating capacity was 26 first class in all units - located in one of the driving vehicles - and 174 standard in Class 220, and 162 or 224 in Class 221, depending on unit length. The first Class 220 unit arrived in the UK from Belgium in January 2001, and the units entered service on 21st May that year. Unit 220 018 poses for the camera at Oxford on launch day.

Bottom right. The driver's position in a Class 220 Voyager. The combined power and brake master controller is on the left, with a screen behind to display unit and service status information. The air brake gauge is just to the right of this screen, while the two dials further right are speedometer (top) and brake percentage gauge (bottom). The units are powered by Cummins QSK19-R diesel engines of 750 hp - one per coach - each driving an Alstom alternator. This supplies power to two 470 hp electric traction motors, one per bogie on each vehicle. Maximum speed is 125 mph - the same as the HSTs - but acceleration is significantly better.

Virgin
Oxford
220
018

DRA

The coaches in the Voyager units are identified by letters - A, B, C, D, F in the five-coach sets, and A, C, D, F in the fours. The absence of coach E was to allow for the addition of an extra vehicle to increase capacity, but this never happened. Some internal changes have been made during the units' lives, but as originally delivered, coach A was first-class, together with the driving cab, train manager's office and a small catering area to provide at-seat service to passengers in that coach. Coaches B, C and F were standard-class, the latter with the driving cab and a bicycle storage area at the outer end. Coach D was also standard class, but with a self-service shop to provide catering and other goods to standard class passengers, similar to the facility found on the West Coast Pendolinos. This view shows the shop in coach D of unit 220 007. There is a microwave oven to the left of the steward, and equipment for dispensing hot drinks behind him.

# BR Network SouthEast sector

**Above. The London and South East sector was created on 4th January 1982, taking on responsibility for passenger services throughout the region, including the intensively used commuter routes serving the Capital. The existing BR business units remained for the time being, with day-to-day operations continuing to be handled by the regions. Greater change came in June 1986 with the re-branding of the sector as Network SouthEast. Superficial signs of this were apparent in new signage and colour schemes, but there were major structural and organisational changes as well, with the sector acquiring considerable autonomy in the areas of timetabling, marketing, rolling stock specification and infrastructure improvements. To passengers, one of the most obvious outward signs of the change was the new livery applied to rolling stock, comprising blue upper and grey lower body panels, separated by red and white stripes with an upsweep at the ends of locomotives and multiple-units. The early version of the livery and branding - a darker shade of blue was used later - is seen here on a Class 319 unit.**

**Below. One of the major infrastructure improvements to reach fruition during the Network SouthEast era was electrification of the line from Branksome to Weymouth, completed in 1988. This filled in the gap left at the time of the Bournemouth electrification scheme in 1967, and brought an end to the changeover from electric to diesel traction at Bournemouth, which had been a feature of the service for 21 years. In addition to the extension of the third rail, brand new rolling stock was provided to replace the Mark 1 based REP and TC units, which had not exactly been 'new' on their introduction in 1967 - most of the vehicles being refurbished locomotive-hauled coaches, some dating from 1951. The new stock comprised 24 five-coach Class 442 units formed of 23 metre main-line Mark 3 coaches. The end vehicles had stylish, streamlined cab fronts and windscreens, as seen in this view of one of the units setting out on a westbound trip from Waterloo. The powered centre coach contained guard's and luggage accommodation and a small buffet area, and the 442s brought, for the first time in a Southern Electric unit, the air-conditioned standards of comfort that InterCity travellers had been enjoying for more than a decade. One aspect of the Southern's 'make do and mend' policy continued, however - the 442s' traction motors were recovered from the withdrawn 4-REP units!**

Above. Another major enhancement completed in 1988 was the opening of the 'Thameslink' route, allowing direct services to operate between the former Southern Region and the Midland Main Line. With the numerous connections available en route, this meant that passengers from the south and south east could travel to the Midlands and the north without the need to transfer between London termini. The major infrastructure works involved the reopening of the Snow Hill Tunnel between Holborn Viaduct and Farringdon - which had last been used in 1970 - at a cost of £1.4 million. New rolling stock was also provided in the form of Class 319 units - a development of the existing Mark 3 based Class 317s - but with dual-voltage capability to allow operation from the 750 volts DC third rail south of the Thames, and 25 kv AC overhead to the north. Changeover between the two systems took place during the scheduled stop at Farringdon station, with the pantograph raised on northbound services and lowered on southbound ones. The 319s were built by BREL at York in two batches - 60 units of Class 319/0 between 1987 and 1988, and a further 26 of 319/1 in 1990. The main difference between the two sub-classes was that 319/0 had standard class only accommodation, whereas 319/1 had 16 first class seats, and were concentrated on the longest distance services between Brighton and Bedford. In this view, second batch unit 319 163 negotiates the approaches to London Bridge station with a Bedford-bound service, pursued by some traditional Southern Region EPB units, the latter approaching the end of their working lives.

Top right. One major task facing Network SouthEast in its early days was the replacement of large fleets of rolling stock approaching life expiry. As mentioned previously, the Weymouth electrification included a small fleet of dedicated new units, while the Thameslink project saw the introduction of the Class 319 fleet. By the end of the 1980s, the 'Kent Link' suburban services were still operated by a fleet of around 200 slam-door EPB units dating from the 1950s, and replacement of these was considered a priority. Orders were placed with BREL and Metro-Cammell for 147 four- and 43 two-coach 'Networker' units, designated Class 465 and 466 respectively, with the first entering service in December 1992. The Networkers shared some attributes with their predecessors - high-density accommodation in a two-plus-three seating layout and a 75 mph maximum speed - but there the similarity ended. Entry and exit was by double sliding/plug doors at the one third/two thirds positions, and traction power was provided by three-phase AC motors, fed from the DC live rail via solid state converters. Seating capacity of a four-coach set was 348 against a maximum of 368 in a 4-EPB unit, and there was no guard's/luggage accommodation, but one improvement - and an innovation on suburban stock - was the provision of a single toilet in each unit. Here, a Class 465/2 unit nears completion at Metro-Cammell's Washwood Heath Works in Birmingham.

Right. Another Network SouthEast success was the upgrading of the Chiltern route from Marylebone to Aylesbury and Banbury, under the first scheme to carry the description 'total route modernisation'. As the term implies, this covered complete resignalling, refurbishment of stations, and provision of new rolling stock, giving the passengers the impression that they were getting a completely new railway. Work started in 1988 and was completed in 1991, at a total cost of £70 million. The resignalling element saw the closure of a number of mechanical signal boxes and the replacement of traditional semaphore signals by colour-lights controlled from a new signalling centre at Marylebone. This view, at Princes Risborough in the chilly winter of February 1991, shows a fine pair of Western Region lower quadrant signals along with their soon-to-be-commissioned replacement. Princes Risborough North signal box, in the background, was closed soon after this photograph was taken, and gradually fell into a derelict state. However, it has since been restored to a high standard thanks to the hard work of the Chinnor and Princes Risborough Railway Association.

Left. The Networker programme had transformed the Kent Link suburban lines, with new stock replacing the old EPB units, but this still left the longer distance services to the Kent Coast to be addressed. These were worked mainly by a large fleet of 4-CEP units dating from 1956-62. They had benefitted from a mid-life refurbishment in the early 1980s, but in the normal course of events would become due for withdrawal by the end of the century. In addition, recommendations following the 1988 Clapham Junction accident envisaged the replacement or significant modification of all Mark 1 vehicles within a similar timescale. The initial replacement proposal was for a Class 471 'Networker Express' unit, and a mock-up of the front end was displayed at London's Victoria station during 1991. This view shows the mock-up at Crewe, before being unveiled to the public. As can be seen, it bears a close resemblance to the suburban Networker cab, but with a gangway connection allowing through access between units, as on the stock it would replace.

Top left. New rolling stock for the Chiltern modernisation scheme came in the form of Class 165/0 'Network Turbo' diesel multiple-units, built by BREL in York from 1990 to 1992. This sub-class comprised 39 units, 28 in two-coach formation, with the remainder being 3-coach. All vehicles were powered by a single Perkins 2006-TWH diesel engine rated at 350 hp, driving the wheels through a Voith hydraulic transmission. Structurally and visually, they were very similar to the electric 'Networker' units built for the Kent Link routes, sharing the same aluminium bodyshells. Taking advantage of the more generous loading gauge on the Chiltern and Thames Valley lines, though, the coaches were three metres longer and three centimetres wider than their electric cousins. Follow-on orders up to 1993 saw further fleets of Class 165/1 and Class 166 units built for main-line services out of Paddington, both having 90 mph capability against the 75 mph maximum of the 165/0s, with the 166s boasting air-conditioning as well. The Chiltern sets were based in a new, purpose-built depot at Aylesbury, seen here during its opening ceremony on 14th May 1991.

Below. Standard Class accommodation in the Class 471s would have been in a two-plus-three layout, though with high-backed seats providing some additional comfort. However, first class passengers in each four-coach unit were to be accommodated in three six-seat compartments accessed by a side corridor, as seen in this mock-up. Such internal layouts were almost unknown in new stock by this time, although the Class 442 sets built for the Weymouth electrification a few years earlier had them. In the event, the combined effects of the early 1990s recession and the break-up of Network SouthEast as part of the privatisation process resulted in the Class 471 being quietly abandoned. A small build of 41 Class 365 units – a dual voltage, 100 mph variant of the suburban Networker – was produced by York Works before it closed in 1996, but only 16 of these found their way to Southern lines, the rest exploiting their 25 kv capability on the Great Northern routes out of King's Cross. The main lines in Kent had to wait until the 21st century before receiving significant fleets of new stock, and the last 4-CEP units soldiered on until 2005!

**Below. Following previous mention of the abortive Class 471 plans, it seems only right to include a view of the venerable 4-CEP units that they were intended to replace. Here, unit 1503 gets away from a snowy Hildenborough with a London-bound service. The date is not recorded, but it looks very smart, possibly in ex-works condition. This unit started life as 7101, the very first of the prototype sets, introduced in February 1956. Along with the other prototypes, it was allocated to the Central Division of the Southern Region, and remained there until going to Swindon for mid-life refurbishment in 1982. It was released from there renumbered into the 15xx series, and then moved to Ramsgate for use on Kent Coast services. All of the 1956-built units were withdrawn during 1993, after 37 years' service, and so narrowly missed surviving into the privatisation era.**

**Top right. Stansted Airport, in Essex, was connected to the railway system during Network SouthEast's tenure. Approval for the link was granted in 1986, and the branch into the airport, built by British Rail at a cost of £40 million, opened in 1991. Construction of the three-platform station under the terminal building was funded by the British Airports Authority (BAA). A half-hourly service was provided between London Liverpool Street and the airport, operated by a dedicated fleet of five Class 322 electric multiple-units. These were a variant of the Class 321s widely used on both Great Eastern and West Coast main line services, the main differences being a larger first class area, and low density two-plus-two seating in standard class. Externally, they wore a hybrid BAA/Network SouthEast livery, as seen here on unit 322 482, bringing up the rear of an airport-bound service. The branch leaves the London to Cambridge main line via a triangular junction between Stansted Mountfitchet and Elsenham stations, and passes under the runway via Stansted Airport tunnel to reach the terminal on the eastern side. The northern chord of the triangle is used by services to Birmingham New Street and Norwich.**

**Bottom right. Although the Network SouthEast era was notable for its service and infrastructure enhancements, there were some station closures as well. One such was Tilbury Riverside, which closed on 30th November 1992. This view shows it in a fairly run down state just a few days before, on 25th November, with unit 312 794 awaiting departure time with a service to Upminster. Opened in 1854, and once boasting five platforms, the station served a nearby pier from which a ferry service operated across the Thames to Gravesend (and continues to do so). It was accessed via a triangular junction between Tilbury Town and East Tilbury stations, and for much of its life was served by trains from Fenchurch Street to Southend and Shoeburyness via Rainham, which reversed there *en route*. From 1986, few through services called, replaced by a shuttle service from Upminster. Its decline and eventual demise arose largely from the development of the Dartford Crossing - both the original tunnels and the later bridge - which reduced the importance of the Gravesend ferry for river crossings. In its last years, the revenue from the train service was dwarfed by its operating costs, making closure inevitable.**

50
322482
Stansted Airport Tunnel
Tye Green Portal
Tunnel length 1795 metres

The influx of new diesel and electric multiple-units as detailed previously allowed the sector to end all locomotive-haulage of trains by 1993. This brought huge benefits to passengers and operators alike, as much of the existing rolling stock was approaching life expiry, and proving increasingly difficult to keep running reliably. One of the last routes to change over was that from Waterloo to Exeter, which had been operated by Class 50 locomotives hauling rakes of non air-conditioned Mark 2 coaches from around 1980. Class 47s began to take over from the 50s from 1990, as reliability of the latter deteriorated, but a longer term solution was required. In the event, the economic downturn meant that the Regional Railways sector found that it had 'over ordered' the Class 158 'Express Sprinter' units, and the surplus was offered to Network SouthEast. With modification work carried out by Babcock Rail at Rosyth Dockyard - including the provision of first class accommodation - these became Class 159 units, and were gradually introduced on the route from 1993. In the last days of Class 47 operation, locomotive 47 708, *Templecombe*, heads a Waterloo to Exeter service near East Coker, just west of Yeovil Junction. The locomotive was withdrawn from service on 1st April 1995, and scrapped in August the same year.

Above. The Waterloo and City Line has long been one of London's railway idiosyncrasies. Opened in 1898 to link Waterloo station with the City, it is the shortest underground line in the Capital, at around one and a half miles, and completely unconnected from the rest of the tube network. Its other distinguishing feature is that it was operated by British Rail and its predecessors until April 1994, when it was transferred from Network SouthEast to London Underground Ltd. The original rolling stock was of American origin, shipped over to the UK and assembled at Eastleigh works. After 40 years in service, these vehicles were approaching life expiry, and so new stock - with a distinctive Art Deco design - was ordered from English Electric and delivered in 1940. This comprised twelve motor coaches - with cabs at each end to allow single operation during quiet periods - and sixteen trailers which were marshalled between pairs of motors to form trains of up to five coaches. These vehicles were originally designated Class 453, later becoming Class 487. This view, at Bank station during the Network SouthEast era, shows motor coach No. 53 arriving with a service from Waterloo. The 487s were replaced after more than five decades in service by London Underground 1992 Stock sets, introduced in 1993.

Left. One or two non-standard liveries were applied to the sector's rolling stock over the years, including the 'Stansted Express' colours covered earlier. Another was this attractive yellow and blue scheme, worn by unit 321 334, seen here on the Great Eastern main line. This was applied by Ilford depot to commemorate the twinning of London Liverpool Street and Amsterdam Centraal stations on 2nd December 1993, and is a version of the livery carried by Dutch Railways (Nederlandse Spoorwegen - NS) rolling stock. NS logos were also applied to the unit (one is just visible at the inner end of the rear coach), together with the name, Amsterdam. The two stations had long had a connection as the British and Dutch termini of the rail and ferry service via Harwich and the Hook of Holland.

Shoeburyness
302 225
2

Top left. Services on the Great Eastern main line between London, Clacton, and Walton-on-the-Naze were operated by Class 309 units from their introduction in 1962 until final withdrawal from the route in 1994. Utilising the Mark 1 coach design, these units were built to full main line standards, with first and second class accommodation in a mixture of compartment and open saloon layouts, and griddle cars in some sets. They were the first BR electric multiple-units to be capable of 100 mph running. Between 1985 and 1987, the entire fleet underwent a refurbishment programme, which saw compartment accommodation retained in first class only, new seating and interior panelling throughout, and double-glazed 'hopper' windows replacing the original sliding vent type. The griddle cars were withdrawn and replaced by ex-locomotive hauled coaches, refurbished to the same standards as the rest of the units. The same period also saw a small increase in the units' areas of operation, with continuing electrification work allowing them to work some services to Harwich, Ipswich and Norwich. Most of the 309s were scrapped following their withdrawal from the Great Eastern lines, but seven units saw further service around the Manchester area until the end of the century. This view shows a first class compartment in a refurbished unit, the new hopper window being clearly visible.

Bottom left. Evening rush-hour at London Fenchurch Street. Class 302 unit 302 225 rolls into the station prior to forming a service to Shoeburyness, while crowds of home going commuters wait in hopeful anticipation of getting a seat! The 302s had a long association with the London, Tilbury and Southend line, working services along it from 1962 until final withdrawal in 1998. These four-coach units were outer suburban in layout - largely second class, but with first class accommodation in one vehicle and toilets in two. Withdrawals started in 1984 when newer Class 310 units began operating on the route. The 302s that remained underwent a refurbishment programme which saw the first class section moved from an intermediate vehicle to a driving coach, the original Gresley bogies replaced by the B4/B5 type on trailer vehicles, and gangways installed between coaches within each set. The original 112-strong fleet had been reduced to just 30 units by 1992, with these remaining in service into the privatisation era.

Below. Class 166 unit 166 211 passes the picturesque setting of Little Bedwyn, with the village and church on the left and the Kennet and Avon Canal to the right, on a May morning in 1996. The unit is operating a service from Paddington to Bedwyn, one of the workings for which this 90 mph, air-conditioned variant of the 'Network Turbo' family was specifically designed. The train's destination station, set in the village of Great Bedwyn but providing a railhead for nearby Marlborough, has long been a terminating point for services from London, Reading or Newbury. For many years, passengers wishing to travel further west had to backtrack to Reading or Newbury first, but some through services do now operate. The Network SouthEast era was drawing to a close on this route when this picture was taken, with the new Thames Trains franchise commencing operations just five months later.

317660
50
WAGN

Stansted Skytrain
EXPRESS RAIL LINK
322485

Top left. As with InterCity, the Network SouthEast sector was split up into Train Operating Units (TOU) from 1994, ready to be let as franchises to private bidders. The Great Northern and West Anglia TOU, covering the routes from King's Cross and Liverpool Street to Peterborough, Cambridge and King's Lynn together with the associated London suburban lines, was let as the West Anglia Great Northern (WAGN) franchise from 5th January 1997. The new operator was Prism Rail, a company specifically formed to bid for the new rail franchises, and run by individuals with previous experience in bus operations. WAGN took on the rolling stock fleet formerly operated by British Rail over these routes, comprising electric multiple-units of Classes 313, 315, 317, 322 and 365. The Class 317 units underwent a significant internal refurbishment to make them more suitable for longer distance journeys, including new lower density seating and carpets throughout. At the same time, they received a smart new livery featuring white upper and blue lower bodysides separated by a yellow stripe, red doors, and a dark grey 'upsweep' at unit ends. Unit 317 660 is seen wearing these colours as it passes Stratford with a 'special' service.

Bottom left. WAGN took on the former Network SouthEast Stansted Express service, but rebranded it as 'Stansted Skytrain' complete with a new livery of light grey with a yellow stripe. Unit 322 485 shows off this branding as it awaits its next working to the airport at Liverpool Street station. As traffic to Stansted increased, the five dedicated Class 322 units proved inadequate for the demand, and they were replaced on the route from 2000 by nine Class 317 units, specially refurbished for the purpose and reclassified as sub-class 317/7. The 'Stansted Express' branding was restored at the same time. The 322s became 'common user' for a short period, working throughout the WAGN network, before being transferred to ScotRail in 2001.

Above. The Class 365 units were developed during the early 1990s as a dual-voltage, long distance variant of the suburban 'Networker' sets. Internally, the main differences were the provision of small first class areas behind each driving cab, a standard and accessible toilet in each unit, and lower density two-plus-two seating in standard class. In terms of performance, the 365s had 100 mph capability against the 75 mph maximum of their suburban cousins. A total of 41 units were turned out by York's Holgate Road Works during 1994/5, the last rolling stock to be built there before it closed in 1996. Units 365 501-16 went to Ramsgate depot, to work Connex South Eastern services in Kent, while the remaining 25 were destined for the Great Northern line, and allocated to Hornsey. 'Dual voltage' is a phrase that has to be used carefully with these units, as the Great Northern examples had no shoegear, and so could not draw current from the live rail, while the South Eastern sets had their pantograph wells plated over, and were thus incapable of operating from the catenary! Units 365 517-41 entered service in December 1996, shortly before WAGN started operations, and so were delivered in Network SouthEast livery, which they continued to show for many years afterwards. Unit 365 517 wears these colours as it sets off from King's Cross into Gasworks Tunnel.

**Above. A Stansted Airport-bound passenger appears to be looking for space on board unit 317 655 at Liverpool Street, while the guard and dispatcher seem keen to get the train away - perhaps the scheduled departure time was 14.30? The picture is undated, but was presumably taken during the late 1990s, when the Class 322 units were struggling to provide sufficient capacity, but had not yet been replaced by the dedicated Class 317/7s. The unit shown here wears the smart WAGN livery introduced shortly after the start of the franchise, but many traces of the Network SouthEast era can still be seen; the digital clock, the red painted notice board stand underneath, and - just visible at the far left - a sister unit in the old sector's colours.**

**Below. A bucolic scene at Offord D'Arcy in Cambridgeshire, between St Neots and Huntingdon on the East Coast main line. Class 365 unit 365 538 heads north between the River Great Ouse and the village church with a WAGN service from King's Cross to Peterborough. On 10th May 2002 a sister unit, 365 526, was involved in a serious accident at Potters Bar, some 40 miles further south, while working a King's Cross to King's Lynn service. Travelling permissibly at just under 100 mph, the third and fourth coaches became derailed when a set of facing points just south of the station moved underneath the train as it was passing over. The rear coach broke away from the rest of the train, turned broadside on to the direction of travel and rolled through 360 degrees before coming to rest lying across both platforms and wedged under the canopies. Six people travelling in the rear coach were killed, as was a pedestrian under a nearby bridge, hit by falling masonry dislodged from the parapet. The front three coaches of the train remained on the rails and came to a stand a short distance north of the station. Subsequent investigation found that the points had failed due to poor maintenance, and legal action was taken against Network Rail and its maintenance contractor.**

Busy evening rush-hour scene at Liverpool Street. A variety of rolling stock is in view as commuters stream along the platforms to board homeward bound trains. At the far left is a Norwich service, formed of Mark 2 coaches still wearing InterCity colours. Next to that, a Class 321 unit waits to depart with a shorter working down the Great Eastern main line, possibly to Southend Victoria, Clacton, or Harwich. Nearest the camera, a Class 322 unit has arrived with a service from Stansted Airport, while a pair of Class 317s prepare to set off towards Cambridge from platform 5.

# First Great Eastern

Above. The Great Eastern franchise was awarded to First Group, with operations commencing in January 1997. With Liverpool Street as its London terminus, the company operated services to Southend and Southminster, down the main line to Colchester and Ipswich, and the branches to Clacton, Walton, Harwich, Sudbury and Braintree. The network was largely electrified, with the exception of the Sudbury branch, and most services were in the hands of former BR multiple-units of Classes 312, 315 and 321. The shuttle service between Marks Tey and Sudbury was worked by diesel multiple-units, hired in variously from Silverlink and Anglia Railways. This view, taken before the operator's new corporate identity had been fully developed, shows a Class 321 unit wearing 'Great Eastern Railway' branding on top of Network SouthEast colours.

Top right. Wickford station lies on the line between Shenfield, on the Great Eastern main line, and Southend Victoria. As well as serving the local town, it is also the junction for the branch line to Southminster. All trains between Southend and London call here, and for most of the day the station is the terminus for the shuttle service from Southminster, although some trains on the branch run to or from Shenfield and London during peak times. This view shows Class 321 unit 321 337 calling with a Southend service. The 321s were built between 1988 and 1991 in three sub-classes: 321/3, 66 units for Great Eastern services; 321/4, 48 units for the West Coast main line; and 321/9, three units for Regional Railways to work West Yorkshire Passenger Transport Executive services between Doncaster and Leeds. One of the driving vehicles was equipped with a lockable compartment behind the cab, which could be used for conveying mail. These vehicles were identified by the suffix 'P' to the unit number, which can be seen here.

Bottom right. A close-up view of the First Great Eastern livery as applied to one of the operator's Class 321 units. It comprised bodyside panels of grey, green and blue, tapering to a point or 'downsweep' at the unit ends. This is a driving trailer composite open (DTCO) vehicle, with twelve first class seats in the section between the cab and the first set of doors. This vehicle was marshalled at the opposite end of the unit to the standard class DTSO, which contained the lockable mail area. It is interesting to note that unlike a number of the private operators, First Great Eastern retained the yellow stripe at cantrail level to denote first class accommodation, and also added a yellow 'block' on the vehicle sides forward of the driver's door, presumably to aid identification of this accommodation as the train approached.

2
Great Eastern
321337-P

First Great Eastern
1

# LTS Rail

Above. The London, Tilbury and Southend lines comprised the two routes from London's Fenchurch Street station to Shoeburyness, via Basildon and Tilbury, together with the branch from Upminster to Grays. The franchise was let to Prism Rail (who also operated WAGN) from 26th May 1996, trading under the name LTS Rail. The company started operations with a fleet of ex-BR Class 302, 310 and 312 units, dating variously from 1959, 1965 and 1976. It set out to replace these with an initial order to Adtranz for 44 Class 357 'Electrostar' units in 1997. However, deliveries of these would not start until 1999, and so as a stopgap LTS hired in 18 Class 317 units from its sister company WAGN. A number of these were repainted into a modified version of the Network SouthEast colours, the red stripe being replaced by a green one, and both this and the upper blue area extended to the unit ends. Unit 317 317 was wearing this hybrid scheme when caught by the camera at Fenchurch Street. As with the Class 321 units described earlier, some of the 317s also had lockable mail areas at one end of the unit, denoted by the suffix 'P' to the unit number, as seen here.

Top right. The South Eastern franchise, covering all of the routes into Kent, as well as the Tonbridge to Hastings line, was awarded to the French-owned operator Connex from 13th October 1996. Although the Kent Link suburban services were in the hands of modern rolling stock thanks to the Networker programme of the early 1990s, the operator inherited a significant fleet of Mark 1 based multiple-units which continued to operate the main line services to the coast. As mentioned earlier, some small inroads were made into this fleet by Class 365 'Networker Express' units, with 16 of these allocated to Ramsgate depot, and the remaining 25 to Hornsey for Great Northern services out of King's Cross. After a lengthy commissioning period, which started under Network SouthEast and was completed some months into Connex's tenure, the Ramsgate-based units finally entered service on 16th June 1997. They were deployed mainly on Victoria-Medway Towns-Ramsgate services, but also operated over the route from Charing Cross to Margate via Maidstone East, Ashford and Canterbury West. On 9 July, units 365 510/507 pass Sole Street with one of the former workings, the 11.05 Victoria to Ramsgate service. One consequence of the units' protracted entry to service was that all 16 were delivered - and undertook much of their test running - in Network SouthEast colours, but had acquired this early version of the Connex livery by the time they were commissioned. This retained the blue upper body panel from the Network SouthEast scheme, but replaced the lower white, red and grey stripes with Connex yellow. (Author).

# Connex South Eastern

Right. In another, minor, attempt at fleet 'modernisation', Connex South Eastern took on twelve Class 508 units from Merseyrail and deployed them on the Medway Valley, Redhill to Tonbridge, and Sittingbourne to Sheerness routes from the winter timetable change in 1998. The Class 508s - a development of the experimental 'PEP' units of the early 1970s - were originally introduced on the Southern Region in 1979, and operated suburban services on the South Western Division. From 1982, they were displaced by newer Class 455 units, reduced to three coaches, and transferred to Merseyside, where they have remained ever since. The move of these twelve sets back to the south-east - facelifted, repainted in Connex livery, and renumbered 508 201-212 - was therefore something of a homecoming! They were, of course, 19 years old at the time of their return, so far from new, although younger than the CEP, CIG and VEP units they replaced. Favourable comparisons ended there, though, and their low-backed seats and lack of toilets were not popular, especially on the lengthy runs from London to Tunbridge Wells, and Maidstone West to Three Bridges. On 8th October 1999, unit 508 208 passes Salfords with one of the latter workings.

The Connex South Eastern franchise was set to run for 14 years, and contained a requirement to replace all of the existing Mark 1 'slam door' rolling stock. The operator met this with an order to Adtranz in Derby (later Bombardier) for new Class 375 'Electrostar' units. These were a variant of the company's 25 kv AC Class 357 unit built for Southend services, itself a development of the 'Turbostar' family of diesel multiple-units of Classes 168 and 170. Structurally, these all comprised welded aluminium side, roof and floor sections bolted together to form the vehicle bodies. The design was modular in concept, with cables and piping pre-installed in the various sections before assembly, and driving cabs also fully fitted out before being slotted into the vehicle ends. Despite the predominant use of aluminium, steel was used in areas subject to higher stresses or requiring greater crashworthiness characteristics, such as coupler assemblies and bogie bolsters, as well as sections of the vehicle ends. Windows were of the continuous glazed 'ribbon' style, rather than individual apertures, and this, combined with sliding/plug external doors, gave the vehicle bodysides a particularly smooth appearance. Externally, the main feature distinguishing the Connex units from the 357s were the unit end gangways, prominently displayed in this view of a completed cab front at Adtranz's Derby Works in September 2000.

Above. The Electrostars entered service in 2001, initially on Chatham line services from Victoria. However, a software 'glitch' soon became apparent, which resulted in the on-board computers taking an excessive amount of time to 'reboot' after units were coupled or uncoupled. This was a major problem, as most of the workings on which the units were rostered divided at Faversham to form portions for Ramsgate and Dover, with a balancing attachment being made in the up direction. As a temporary measure, a timetable from some years earlier was reinstated, which saw separate fast services running from Victoria to Ramsgate and Dover on alternate half hours, with slow services to each destination making a cross-platform connection at Faversham. The software problems were finally resolved in 2003, allowing the restoration of portion working not just on this route, but on the south eastern main line via Ashford as well. This view at Rochester shows unit 375 703 bringing up the rear of a Victoria-bound service. This unit belongs to sub-class 375/7, broadly similar to the initial 375/6 sub-class, but lacking dual-voltage capability.

Right. Buffet cars had disappeared from Kent's train services by 1982, but Network SouthEast introduced refreshment trolleys on the Hastings line following its electrification in 1986, and these gradually spread to other services. Connex continued to provide these on its longer distance trains, and this view at London Bridge shows a steward loading a trolley into the brake van of a 4-CIG unit - no easy task by the looks of it! The unit itself is in a plain, off-white, unbranded livery. Five 4-CIG units - 1701, 1736, 1737, 1742 and 1748 - were given this treatment in 1998 by lessor Angel Trains, to signify that they were intended for use on 'short notice' or 'low mileage' leases. In the event, these five lasted in service until 2003/4, the final three remaining allocated to Ramsgate until the end, and outliving the Connex South Eastern franchise in the process. The operator ran into financial difficulties in 2002, and received a £58 million bail-out from the Strategic Rail Authority (SRA). The end date of the franchise was brought forward to 2006 at the same time, but the position continued to deteriorate and Connex was stripped of the franchise in November 2003, with operation transferring to the publicly owned South Eastern Trains.

# Connex South Central

Above. In addition to the South Eastern franchise, Connex were also awarded the neighbouring South Central one, covering the south London suburban lines, and routes through Surrey and Sussex to the South Coast from Hastings in the east, to Portsmouth in the west. It commenced operations on 26th May 1996, initially trading under the name of the shadow franchise, 'Network SouthCentral', before rebranding as Connex South Central on 13th October. Like its South Eastern counterpart, it adopted a predominantly white and yellow livery with a narrow blue band at solebar level, and this was applied to rolling stock using a painted base colour and yellow vinyl transfers. Some problems were encountered initially with the application of the vinyls to slam door vehicles, and until these were resolved the units concerned ran in plain white - not to be confused with the 'unbranded' treatment applied to some 4-CIG units, as mentioned earlier. 4-VEP unit 3541 looks pristine in white with blue solebar stripe, when captured by the camera at Victoria.

Top right. One innovation introduced by the operator from June 1997 was a direct service between Gatwick Airport and Rugby, providing a useful direct link to the Midlands and other destinations on the West Coast main line. Rugby was chosen as the northern terminating point partly because it had under-used bay platform capacity. The service was operated by Class 319 units, using their dual voltage capability, and running via the West London line between Clapham and Willesden Junctions. It was later extended to Brighton, but was subsequently cut back to Watford Junction at the northern end. On 6th February 1998, unit 319 005 - smartly turned out in Connex livery - awaits departure time with a southbound service at Rugby.

Bottom right. The iconic Ouse Valley Viaduct, near Balcombe, is the setting for this view of a service from London to the south coast on 5th May 1998. The viaduct - designed by the London and Brighton Railway Company's chief engineer John Rastrick - was built between 1839 and 1841, as part of the construction of the line from London to Brighton. It is carried on 37 brick arches, has a length of 1,480 feet, and a maximum height of 96 feet. Approximately eleven million bricks were used in its construction, and the total cost at the time was £38,500, equivalent to well over £3 million today. The train is formed of three 4-CIG units led by 1834, wearing Connex colours, while the second unit is still in the Network SouthEast scheme. The particular working is not recorded, but at this time many services on the route divided at Haywards Heath - about three miles south of this point - with portions then running along the Coastway routes to Hastings in the east, and as far as Bournemouth to the west.

319005

SPAD I
OD 64
39
SPAD I
OD 62

1704
64
1704

# South West Trains

Top left. The line from Oxted to Uckfield originally formed part of a through route to the south coast, until closure of the section between Uckfield and Lewes in 1969. Although the route through Oxted to East Grinstead was electrified in 1987, the business case could not be made for electrification to Uckfield, and the branch remains diesel operated today. Further rationalisation followed in 1991 when the line was singled south of Hever, with long passing loops at Ashurst and Crowborough. This singling was an indirect contributory factor to the head-on collision that occurred between two trains between Ashurst and Cowden on 15th October 1994, sadly resulting in the deaths of three railway staff and two passengers. Following the subsequent inquiry and its recommendations, SPAD (signal passed at danger) indicators were installed in conjunction with all of the signals protecting entry to the single line sections. These comprise three red lights positioned a short distance beyond the relevant signals. Should a train pass one of these signals at red, the top and bottom lights in the indicator will flash, while the centre light shows a steady red aspect. Two such indicators are seen here at Crowborough on 24th May 1998, while Class 205 unit 205 028 leads a sister unit off the single line section with an Uckfield-bound service.

Bottom left. The 'East Coastway' route, between Brighton and Hastings, is depicted in this view near Glynde. 4-CIG unit 1704 leads another of the same class towards Lewes with a Victoria-bound service. One of these units probably originated from Hastings, coupling to another at Eastbourne to form an eight-coach train for the run to London. A graffiti artist appears to have been at work on the front of unit 1704, a problem which was sadly to afflict much of this stock in its later years. There is also some evidence of minor subsidence at the sides of the low embankment behind the train, with both colour-light signals straying a little from the perpendicular!

Above. The South West Trains franchise, operating suburban routes in south west London and longer distance services to Portsmouth, Southampton, Bournemouth, Weymouth and Exeter, was awarded to the Stagecoach Group in 1995. Scotland based Stagecoach was best known as a bus and coach operator, but had made a brief foray into rail operations between 1992 and 1993 with 'Stagecoach Rail', providing overnight seated services between London and Aberdeen, using Mark 2 coaches hired from British Rail attached to the scheduled sleeper service. Its subsequent South West Trains franchise represented one of the first to commence operations, along with Great Western, on 4th February 1996. Although new liveries were gradually applied to rolling stock, most initially remained in the Network SouthEast colour scheme, with suitable branding applied, as seen here on Class 159 unit 159 016 passing Battledown flyover, west of Basingstoke, with an Exeter to Waterloo service. The branding, in the form of 'SWT' stickers, can just be seen below the windows at the mid-point of each coach.

**Right. A close-up view of Driving Motor Standard (DMS) vehicle 57875 of Class 159 unit 159 003, highlighting two points of interest. The first is the prominent legend on the white band beneath the windows, indicating ownership of the franchise in addition to the SWT branding applied elsewhere on the vehicle. The operation got off to a bit of a shaky start, with many drivers being offered a generous voluntary redundancy package in an effort to cut costs. Unfortunately, it then transpired that the full service could not be run without significant overtime working, and a period of regular cancellations followed, ultimately resolved by the recruitment of additional drivers! There was also some controversy around Stagecoach's acquisition of the rolling stock leasing company Porterbrook, in what some considered a poor deal for the taxpayer. The potential reputational damage to the operation resulting from these events may have been a factor in the decision to remove any reference to the company from later versions of the livery, leaving just South West Trains as a standalone brand. On a lighter note, the vehicle also carries the name, *Templecombe*, commemorating the station of the same name between Salisbury and Exeter, closed in 1966 and reopened in 1983.**

**Below. The first version of the South West Trains livery is seen here in this view of Class 455 unit 5714 standing at Waterloo station. Essentially, it was a variation of the Network SouthEast scheme, retaining the large blue area around the windows, but replacing the grey, red and white stripes with white, orange and red to reflect the Stagecoach group's house colours. The livery was applied in secret to a single Class 442 unit - 2402 - which was then revealed on launch day at Waterloo, but gradually began to appear across the rest of the company's fleet over the following months.**

Concurrent with a major refurbishment programme in 1998, a new livery began to appear on the operator's 'flagship' Class 442 units, which operated most services to Bournemouth and Weymouth, and some to Portsmouth. Vehicle bodies were predominantly white, relieved by blue bands along the windows and lower bodysides. The red and orange colours formed an upsweep - or 'swoosh' - at each end of the unit, starting at solebar level and moving up and over the roof about halfway along the vehicle, and orange lining was applied along the top of the lower blue bodyside band. A new logo, comprising the 'South West Trains' name in bold blue and orange capitals, with a dot over the 'I', was applied to the bodysides above the windows. Internally, the refurbishment work included a reduction in the size of the van area to provide additional seating, and the provision of stowage racks for five bicycles. The new colours sat particularly well on these units, as seen in this view of a pair descending from Battledown flyover with a Waterloo-bound service.

Below. In the early days of privatisation, station nameboards were 'customised' to fit individual operator's branding and house styles, gradually dispensing with the once ubiquitous 'Rail Alphabet', introduced as part of British Rail's new corporate identity from 1965. This led to a huge variety of colour schemes and font styles at stations across the network, some - it has to be said - rather clearer than others. One of South West Trains' early examples is seen here at Clapham Junction. It has the red and orange 'swoosh' which formed part of the revised livery on the Class 442 units, but the lettering - while in blue - is still in Rail Alphabet; later examples were to use a slightly modified font style. One of the operator's 4-CEP units can just be seen passing behind, its identity revealed by the Commonwealth bogies under the vehicles.

Top right. As a first step towards replacing the significant fleet of Mark 1 based multiple-units that it had inherited from British Rail, South West Trains, together with leasing company Porterbrook, placed an order with Alstom in 1997 for 30 four-coach units designated Class 458. Intended for outer suburban and main-line stopping services, the 458s had standard class accommodation in a two-plus-three layout, with small first class compartments behind each driver's cab containing two-plus-two seating. Both driving vehicles and one intermediate coach were powered, each having a single motor bogie, giving a total power output of 2,170 hp and a maximum speed of 100 mph. The first unit was accepted for testing in October 1998, and service entry eventually took place in February 2000, but the fleet was beset by problems, both electrical and mechanical. A particular issue concerned the inter-unit gangways, which were concealed behind doors when not in use, and had to be manually extended and connected, a far cry from the much simpler process of coupling Mark 1 units with 'Pullman' gangways and buckeye couplers! As a result, the units generally ran permanently coupled in pairs, but the other continuing problems resulted in them being taken out of service and stored in 2005, until the various issues were resolved. They began returning to service from 2006, by which time the operator had placed its significant order for Siemens 'Desiro' units of Classes 444 and 450. In this view at Waterloo, unit 8025 leads a sister unit out of the station on to the down main slow line, while a couple of 'Eurostar' sets rest at the international platforms in the foreground. As can be seen, the 458s wore the new livery first seen on the Class 442 units, but with a somewhat shortened upsweep at the unit ends.

Bottom right. Another new rolling stock acquisition for the company came in the form of eight Class 170 diesel multiple-units, ordered from Adtranz for delivery in 2000. Full details of this ubiquitous class have been provided earlier under the sections covering the Anglia and Midland Main Line franchises. The South West Trains sets were in two-coach formation, both composite vehicles with nine first class seats in each, 53 standard class seats in one, and 43 in the other which also contained the accessible toilet and wheelchair spaces. They were used to supplement the Class 159 units on services between Brighton and Reading, Waterloo and Salisbury, and occasionally Waterloo and Exeter. However, their lack of end gangways was an operating inconvenience, and in 2006 they were transferred to First Group for use on TransPennine Express services. In exchange, South West Trains took on eleven Class 158 units, which could interwork happily with their existing Class 159 fleet. During its period with the operator, Class 170 unit 170 301 pauses at Basingstoke with a Waterloo-bound service. A Thames Trains Class 165 unit waits in the adjacent bay platform with a service for Basingstoke.

eurostar
eurostar

SOUTH WEST
170 301

# Island Line

**Above. Following many line closures between 1952 and 1966, the Isle of Wight was left with just the eight and a half mile stretch between Ryde Pier Head and Shanklin, this being electrified in 1967. Loading gauge restrictions meant that standard electric units could not be used, and so former London Underground tube stock, suitably modified, was employed instead. The first units used from 1967 had originally been built between 1921 and 1923, and so were over 40 years old when they started work on the island. They remained in service until 1992, when they were replaced by Metro-Cammell units which had first entered service in London in 1938. In their refurbished, two-coach formation, they were designated Class 483. Unique among the passenger franchises at privatisation, the Island Line required the operator to maintain the infrastructure as well as operating the train service. It was awarded to the Stagecoach Group in 1996, who also operated the neighbouring South West Trains franchise on the mainland (and the two franchises were subsequently merged on renewal in 2007). In this view, Unit 483 003 - still wearing Network SouthEast colours - heads north from Brading with a service from Shanklin to Ryde Pier Head.**

**Top right. The Thames Trains franchise was awarded to Victory Rail Holdings, a company whose ownership was initially split between Go-Ahead Group (65 per cent), and a team of ex-British Rail managers (35 per cent). Operations commenced on 13th October 1996, and in June 1998 Go-Ahead bought out the remaining 35 per cent to become the sole owner. It operated services out of Paddington along the Great Western main line to Reading (including the branches to Greenford, Slough, Marlow and Henley), thence to Bedwyn on the Berks and Hants route, and to Oxford, Banbury and via the Cotswold line to Worcester and Hereford. It also operated the link between Reading and Gatwick Airport, and it is on one of these latter workings that Class 166 unit 166 211 is seen at Salfords on 8th October 1999, nearing journey's end with the 11.34 from Reading. Just three days earlier, the operator had been in the news for all the wrong reasons, after one of its Paddington to Bedwyn services had collided head-on with a London-bound HST at Ladbroke Grove, tragically resulting in 31 fatalities. The cause of the accident was attributed to a combination of inadequate driver training and poor signal sighting.**

**Bottom right. Thames Trains was fortunate in inheriting an almost new fleet of Class 165 and 166 'Turbo' units, which had displaced life-expired first generation diesel multiple-units and locomotive-hauled formations a few years earlier. At the start of the franchise, all of these remained in Network SouthEast colours, but once the paintwork warranty had expired, a new livery of white and blue - with green highlighting around the doors - was applied, as seen here on unit 166 204 at Hinksey on 7th April 2001. The unit was caught by the camera while working the 08.58 Stratford-upon-Avon to Paddington service.**

Thames Trains
166211

166 204

Below. Another Class 166 unit, 166 205, approaches Paddington on 30th August 2001. The livery varied slightly between Class 166 and Class 165. The latter, which lacked air-conditioning, did not carry the 'Express' branding, while the green area over their doors was circular, rather than the 'upsweep' arrangement seen here. Also prominent in this view is the overhead catenary, originally installed in 1997 as far as Airport Junction, between Hayes and Harlington and West Drayton, in readiness for the start of the Heathrow Express service the following year. The presence of this equipment, and its effect on signal sighting, was considered to be one of a number of contributory factors in the Ladbroke Grove accident referred to earlier.

Top right. The Network SouthEast Thames and Chiltern division was split at privatisation to create two franchises - Thames Trains, already covered, and Chiltern Railways. The latter was awarded to M40 Trains, a management buy-out backed by John Laing Group PLC and 3i Group PLC. Operations started on 21st July 1996, with services running from London's Marylebone station to Aylesbury, High Wycombe, Bicester and Banbury. The franchise was led by Adrian Shooter, a former BR manager who was determined to build on the success of the total route modernisation project carried out by Network SouthEast between 1988 and 1991. A major aspiration was to create a second main line between London and Birmingham, and to operate a half-hourly service over the route. A serious obstacle to this was the 30-mile single track section between Princes Risborough and Aynho Junction, broken only by a passing loop at Bicester. Between 1998 and 2001, this section was redoubled, and subsequent track and signalling improvements were carried out to make as much of the line as possible fit for 100 mph running. At the London end of the route, one of the multiple-units introduced under the original modernisation scheme, Class 165 set 165 013, stands at Marylebone wearing the operator's colours.

Bottom right. In addition to infrastructure improvements, 100 mph running required new rolling stock, with the Class 165/0 units only cleared for 75 mph. This was met by an order to Adtranz for 19 Class 168 'Clubman' sets, which were delivered in a number of batches between 1998 and 2004. In addition to the higher maximum speed, the 168s were suitably appointed internally for the longer distance journeys on which they would be deployed, with carpeting and two-plus-two seating throughout, and two toilets in three-coach sets, with three in the four-coach examples. All vehicles were powered, each by an MTU 422 hp diesel engine driving the wheels through a Voith hydraulic transmission. The first batch, Class 168/0, comprised five units numbered 168 001-005. Having been originally specified in the last days of Network SouthEast, these retained the Networker style front end as used on the Class 165 and 166 units. Unit 168 001 is seen here on a test run prior to service entry at Cossington on the Midland Main Line. These five units were delivered in three-coach formation, later extended to four coaches.

165 013
Chiltern Railways

168001

Above. The second and third batches of Class 168 units, sub-classes 168/1 and 168/2, were delivered in 2000 and 2004 respectively, all but two of them as three-coach sets. These were essentially variants of the Adtranz/Bombardier Class 170 units, and shared the same front end design as the rest of the 'Turbostar' family. A final batch of Class 168/3, two-coach units, were introduced around 2015 to provide additional capacity, although these had started life as Class 170/3, first with South West Trains, and then TransPennine Express. Sidelit by early spring sunshine, unit 168 112 of the second batch starts away from Leamington Spa with a Marylebone-bound service.

Top right. The backbone of the North London Railways franchise was the West Coast main line from Euston to Northampton - with some services continuing to Birmingham - and London suburban workings between Euston and Watford, Willesden Junction and Clapham Junction, Gospel Oak and Barking, and North Woolwich to Richmond. It also operated the St Albans Abbey branch line from Watford Junction, and the Marston Vale line from Bletchley to Bedford. The franchise was awarded to National Express on 7th February 1997, with operations commencing on 3rd March. It only operated under the original name for a few months, though, before adopting the name 'Silverlink' in September of that year. It inherited a mixed rolling stock fleet from British Rail - Class 117 and 121 diesel multiple-units to work the non-electrified Gospel Oak to Barking and Bletchley to Bedford routes, and electric units of Classes 321 and 313 for everything else. One of the latter, unit 313 117, comes off the West London line with a service from Willesden Junction to Clapham Junction. Three of Battersea Power Station's chimneys can be seen in the background - the fourth is obscured by the tower block to the right.

Bottom right. The operator developed two distinct brands for its services - 'Silverlink Metro' for the London suburban workings, and 'Silverlink County' covering the West Coast main line, St Albans Abbey, and Marston Vale routes. The 'Metro' branding can be seen here on unit 313 121, resting at Watford Junction after arrival with a service from Euston. These versatile units had dual voltage capability, able to work from the 750 volts DC third rail supply or 25 kv AC overhead. This feature was particularly useful on Silverlink services, with Euston to Watford being DC throughout, the West Coast main line and St Albans Abbey branch being 25 kv AC (although 313s did not operate widely on these routes), and Willesden to Clapham a mixture of both.

# Silverlink

# Thameslink

Top left. The 'Silverlink County' brand is represented here by Class 321 unit 321 430 passing Roade, near Northampton, while working a Euston-bound service. The leading vehicle is a Driving Trailer Composite Open (DTCO) with 28 first class seats accessed via the first set of doors. This first class section occupied both the area behind the driver's cab, and part of the centre of the coach as well, with a partition and internal door separating it from standard class. This view also shows to advantage the classic West Coast main line catenary equipment, much of it dating from the 1960s, using frames or 'portals' to support the overhead wires above multiple tracks. By contrast, the later East Coast electrification scheme made extensive use of 'headspan' arrangements, in which there is no rigid crossmember between the uprights, but horizontal tensioned wires instead.

Bottom left. As mentioned earlier, the five 'Stansted Express' Class 322 units ceased to be dedicated to airport services from around 2000, and were used widely throughout fellow National Express franchise WAGN's network before being transferred to Scotrail in 2001. In the meantime, some were occasionally sub-leased to Silverlink for use on services between Euston and Birmingham New Street. One such, unit 322 482, is seen here heading north with a mid-morning service at King's Langley, its grey and yellow 'Skytrain' livery making an unusual sight on the West Coast main line.

Above. The franchise to run the cross-London Thameslink services was awarded to Govia, a joint venture company set up in 1996 specifically to bid for rail franchises, with ownership split between two existing transport operators, the Go-Ahead Group (65 per cent) and Keolis (35 per cent). Services started under this new incumbent on 2nd March 1997, with a fleet of Class 319 units taken on from Network SouthEast. As part of the privatisation process, this class had been split, with Thameslink retaining 66 sets and the remaining 20 going to Connex South Central. Of the latter, seven units were converted to Class 319/2 for use on Victoria to Brighton express services. Modifications included an enlarged first class section with 18 seats in a one-plus-two layout, the provision of a 'snug' area with space for a catering trolley, and two-plus-two seating throughout standard class. Examples of both operators' 319s can be seen in this bird's eye view of Brighton station, the unit in the foreground wearing Thameslink's livery of navy blue with a broad yellow stripe along the lower bodyside.

Top left. The Class 319 units taken on by the Thameslink franchise underwent various reformations between 1997 and 1999 to create two further sub-classes, 319/3 and 319/4. The 26 Class 319/1 units were converted to 319/3 at Eastleigh, the work involving the removal of first class seating to create an all standard class layout, and application of the new livery. These units were used on the shorter distance workings between Sutton, Wimbledon and Luton. The Class 319/4s were converted from 40 of the original Class 319/0 sets, in this case by adding a twelve-seat first class section at the outer end of what had been the Driving Trailer Standard Open (DTSO) 'A' vehicle. In addition, some standard class seats were removed to create a two-plus-two layout in the centre of each vehicle, seat covers were replaced, and carpets fitted throughout. This work was carried out by Railcare at Wolverton, and these units were then allocated to the longer distance Bedford to Brighton services. This view shows the interior of a refurbished Class 319/4 vehicle.

Bottom left. On 5th April 1998, a pair of Class 319/4 units speed across John Rastrick's Ouse Valley viaduct with a Bedford to Brighton service. The pantograph for working from the 25 kv overhead supply north of the changeover point at Farringdon can just be seen at the leading end of the third coach. At this time, virtually all 'fast' trains between London and Brighton were formed of Class 319 units, with Connex 319/2s working the express services from Victoria, and Thameslink 319/4s serving King's Cross Thameslink, Farringdon, City Thameslink, Blackfriars and London Bridge on their way through from Bedford. The older Mark 1 units - 4-CIG and 4-VEP - still held sway on the 'Coastway' services, leaving the Brighton line at Keymer Junction for Eastbourne and Hastings, and Hove for Littlehampton, Portsmouth and Southampton.

Below. On the day following the previous picture, unit 319 427 leads an unidentified sister unit across the complex pointwork outside Brighton station as it arrives with a service from Bedford. This unit started life with Network SouthEast as Class 319/0 unit 319 027 in 1988, taking on its later identity following refurbishment between 1997 and 1999. The destination blind at the top of the driver's windscreen (which would not meet current standards for passenger information systems!) reads 'Brighton via Gatwick Airport'. Gatwick, of course, was always an important destination on the Thameslink route, and when the service first started, one train from Bedford each hour terminated there, only alternate workings continuing to Brighton. Service frequency over the whole route increased to half-hourly from 1992, and to four an hour by 1995. The brick building on the right is a relay room, put in as part of the Three Bridges resignalling scheme, which took over control of the area in 1985.

Regional Railways started life in 1982 as British Rail's 'Provincial' sector, responsible for all passenger services throughout the UK not operated by InterCity or Network SouthEast. It was the most heavily subsidised of all the BR sectors, with costs equal to four times its revenue at the time it was formed. The sector took on an eclectic mix of rolling stock, comprising diesel and electric multiple-units and a quantity of locomotive-hauled coaches. Most of this still wore BR blue or blue/grey livery, although some Passenger Transport Executive (PTE) colour schemes and logos had begun to make an appearance. A new livery was developed for the sector - comprising a light blue band along the windows with a darker blue below - and this was gradually applied to stock as overhauls became due. This was later reversed, with a dark blue upper section separated from a broad, light blue band by a white stripe, with a light grey section below, and this was retained - together with 'Regional Railways' branding - when the sector was renamed in 1989. This later treatment is seen here, newly applied to a Class 101 DMU vehicle.

Right. One of the sector's greatest successes was its procurement and introduction of a whole new generation of diesel multiple-units. As well as allowing the withdrawal of life-expired stock, the superior performance of the replacements led to some journey time improvements, while greater reliability - and therefore availability - meant better service frequencies on some rural routes. The new stock arrived in two tranches - the Class 14x 'Pacer' units, based on the original Leyland bus-bodied prototypes of the late 1970s, and the Class 15x 'Sprinter' sets, intended for longer distance routes where enhanced performance and greater passenger comfort were required. The Class 156 'Super Sprinters' were two-coach units built by Metro-Cammell between 1987 and 1989. They were powered by Cummins NT855-R5 diesel engines of 285 hp - one per coach - and had a maximum speed of 75 mph. Internally, they were all standard class with seats in a two-plus-two layout, and they had end gangways which allowed passengers and staff to move throughout a train formed of two or more units. This view shows the interior of a Class 156 vehicle, with the seats covered in the original style Regional Railways upholstery.

Below. For those routes where even better performance and passenger comfort was required - some still worked by locomotive-hauled formations - the Class 158 'Express Sprinter' was developed. These were built by BREL between 1989 and 1992, and featured a number of innovations compared with the earlier Sprinter sets. The bodyshells were constructed from aluminium, giving some weight savings; air-conditioning was installed, the first time this had been provided on any Regional Railways stock, and disc brakes were used rather than the tread type of earlier units. The vast majority were in two-coach formation, although some units were made up to three coaches by the addition of a non-driving centre vehicle. Power was provided by either Cummins or Perkins diesel engines rated at 350 or 400 hp, one per coach, giving a maximum speed of 90 mph. During the construction process, a number of partially completed aluminium bodyshells can be seen here inside BREL's Derby Litchurch Lane Works.

52865

**Left. Another view at Litchurch Lane shows newly completed Class 158 vehicle 52865, from unit 158 865. This coach was a Driving Motor Standard Lavatory (DMSL) 'B', with 68 seats, accessible toilet, space for wheelchairs and a refreshment trolley, and a public telephone, the latter being a particularly useful feature in the days before widespread mobile phone ownership. This vehicle also sports the revised Regional Railways 'Express' livery developed for these units, with buff and dark grey complementing the original light blue, dark blue and light grey scheme. Soon after entry to service, it was found that these units did not always activate track circuits properly, especially during the leaf fall season. This was a potentially dangerous problem, as it would result in the signalling system failing to detect their presence in block sections. The issue arose from a combination of the stock's comparatively light weight and the use of disc brakes, which failed to remove leaf mulch contamination in the same way that tread brakes did. The problem was eventually solved by a combination of scrubbing blocks to remove the contamination, and the fitting of track circuit actuators which fed a high frequency electric current through the wheels to the rails.**

**Above. There is so much to enjoy in this view of Ely Dock Junction, taken shortly before closure of the signal box on 10th April 1992. Centre stage is Norwich-based unit 158 764, probably working a service from Stansted Airport to Liverpool. To the right is the signal box, with the whole scene framed by some splendid - and soon to be displaced - semaphore signals. Those on the far right control trains coming off the line from Bury St Edmunds and Ipswich; the distant arms are for Ely Station North signal box. Ely's magnificent cathedral - often known as the 'Ship of the Fens' - is visible in the left background, while in front of it can be seen a catenary mast for the forthcoming electrification work, completed and commissioned during 1992.**

Unit 158 722 passes the signal box at Tay Bridge South with a service towards Edinburgh. Just visible below the second passenger saloon window is the 'Express' branding that these units carried in Regional Railways service. Also noticeable is the yellow first class stripe above these first two windows. Most of the 158s were standard class only, but those based at Edinburgh's Haymarket depot had a small first class section at one end seating 15 passengers. The city of Dundee can be seen at the far end of the bridge, on the north bank of the Firth of Tay, with the Cairngorm

Mountains behind. This, of course, is the second Tay Bridge, built between 1883 and 1887. It replaced the original 1878 bridge, parallel and some 60 feet downstream, which collapsed during a severe storm on 28th December 1879. A train crossing at the time plunged into the river, with the loss of all 75 people on board. The cause of the disaster was attributed variously to deficiencies in construction and maintenance, and to the high winds blowing at the time.

Blackpool North
150134
REGIONAL RAILWAYS
Dalby

Platform 2
Sprinter
Ivanhoe Line
Special
Information

# Wales and West

Top left. Some stock working services sponsored by the various PTEs was repainted into the house colours of those bodies, while continuing to carry Regional Railways branding. This view, at Derby's Litchurch Lane works, shows a vehicle from Class 150 unit 150 134 wearing the 'new' Greater Manchester PTE livery of dark grey, red, white and light grey, newly applied as part of a refurbishment programme. The earlier PTE livery, used until around 1992, had been orange and brown, matching those used on its buses. The 150s were built by BREL between 1984 and 1987, mainly as two-coach sets, although some had three. They employed the suburban version of the Mark 3 bodyshell also used for several electric classes including 455, 317, 319 and 321, and were produced in two batches. Class 150/1 (50 units) - seen here - were only gangwayed within sets, while Class 150/2 (85 units), had end gangways allowing access between units. Engines and transmissions were identical to those on the later Class 155 and 156 units.

Bottom left. The 'Ivanhoe Line' was the name given to the project to initially re-open three stations on the Midland Main Line in May 1994, served by a new local service between Leicester and Loughborough. It was intended to be Phase One of a larger scheme, which would also see passenger services restored between Leicester and Burton-on-Trent, although that did not proceed at the time. The stations - Syston, Sileby and Barrow-upon-Soar - had all closed in March 1968, and were rebuilt to serve just the slow lines, which had previously been used predominantly for freight. Barrow and Sileby each have two platforms, while Syston has a single one served by trains in both directions. Passenger services commenced on 27th May, but an opening ceremony was held at Barrow-upon-Soar a few days earlier, on 22nd May, attended by various dignitaries including Roger Freeman MP, Minister for Public Transport. The ceremony is seen here, with unit 156 422 carrying a commemorative headboard. The local service was withdrawn in 2005, with the stations served instead by trains running between Leicester, Nottingham and Lincoln.

Above. The South Wales and West Railway franchise covered services throughout Wales and South West England, with some running to the West Midlands and the North West. It was awarded to Prism Rail from 13th October 1996, but was relatively short-lived. National Express took over from Prism in 2000, and the franchise was split up to become Wales and Borders and Wessex Trains the following year. The growing use of vinyl wraps applied to rolling stock enabled it to be particularly innovative with its liveries, developing a number of eye-catching schemes to promote specific lines, areas, and tourist attractions. This view, on the picturesque branch line from Liskeard to Looe, shows unit 150 266, The *Whitley Wonder*, named after Herbert Whitley, the founder of Paignton Zoo and its associated steam railway. The graphics on the second vehicle promote this attraction, while those on the first advertise the westcountry.now website, billed at the time as the official guide to South West England.

Below. Barnstaple was one of the outposts of the operator's network, and the station is seen here with a pair of Class 153 single-coach units at the platform, 153 355 at the rear. Services from here ran down the 'Tarka Line' to Exeter, with some continuing to Exmouth. Judging by the passengers apparently boarding and leaving the train simultaneously in this view, one would imagine the turnround time was quite tight - whether scheduled or due to disruption is not recorded! Barnstaple station - previously Barnstaple Junction - was once an important railway hub in the area, with three through platforms serving routes to Torrington, Taunton and Ilfracombe in addition to the line to Exeter. The Beeching cuts during the 1960s saw these lines close in 1965, 1966 and 1970 respectively, leaving just the link to Exeter remaining. That route is single track, with only two passing loops over its 39 mile length, and as can be seen, the station now manages with just a single platform. The routes to Torrington and Ilfracombe previously continued towards the camera, and the abandoned and overgrown island platform can just be seen on the right
.

Top right. In order to differentiate its frontline express services from its more general workings, Regional Railways launched the 'Alphaline' brand in 1994. It was applied only to services operated by the air-conditioned, 90 mph Class 158 units, which also provided a trolley refreshment service, seat reservations, and an onboard payphone. On privatisation, some operators of former Regional Railways services retained the branding, Wales and West among them. Initially they applied the Alphaline title and distinctive logo over the existing Regional Railways livery, replacing the previous 'Express' wording below the first passenger window behind the driver's cab. Subsequently, they developed a dedicated livery for these units, with silver bodysides, navy blue doors, and a full-height Alphaline logo on the vehicle sides. Unit 158 746 displays this treatment to good effect as it departs southwards from Manchester Piccadilly. The particular working is not recorded, but Wales and West operated services from Manchester to Cardiff, Penzance, and even London Waterloo.

Bottom right. Another of the operator's attractive liveries promoted the Heart of Wessex line, and is seen here on Class 153 unit 153 305, standing at Gloucester on 28th June 2001. It is coupled to a Class 143 'Pacer' unit, still wearing Regional Railways colours. The Class 153s started life as Class 155 two-coach units, built by Leyland Bus during 1987/8 as part of the 'Sprinter' programme. However, passenger numbers on some rural services could be handled adequately by a single railcar type vehicle, a requirement previously met by the first generation Class 121 and 122 units, which were approaching life expiry. BR therefore took the decision to split 35 of the 155s and convert them to 70 single coach units, which could be deployed on such lightly used routes, or used in multiple with two coach sets to boost capacity. The conversion work was carried out by Hunslet-Barclay in Kilmarnock during 1991/2, and involved the creation of a second driving cab at the 'inner' end of each vehicle, where it had previously been coupled to its neighbour. The new cab, seen at the far left of this picture, is considerably smaller than its counterpart, occupying the space formerly used for luggage stacking beyond the end vestibule. 'Heart of Wessex' is the branding applied to the route from Bristol to Weymouth, via Bath, Bradford-on-Avon, Westbury, Frome, Yeovil and Dorchester.

158746
Stop
Look
Listen

THE
HEART
OF
WESSEX
LINE
Through The Heart of Wessex
from Bristol to Weymouth
52305

GREAt SCENIC RAILWAYS Of DEVON AND CORNWALL
www.carfreedaysout.com
tel 01752 233094
Produced in partnership with the Devon and Cornwall Rail Partnership

# Valley Lines

Top left. Another view of Class 158 units in Alphaline livery, this time consecutively numbered units 158 746 and 158 745 heading east past Valley, Anglesey, with a service from Holyhead. The line diverging behind the train - and also just visible to the right of the leading coach - leads to the sidings originally laid in 1962 to allow the transfer of waste from the Wylfa nuclear power station. The power station was decommissioned in 2015, but the sidings also serve a triangle used for turning steam locomotives which have worked charter services to Holyhead. Another interesting feature on this stretch of line, almost two miles further east, are the emergency colour-light signals operated from the control tower at Valley RAF Station. These, together with their associated distants, are normally unlit, but can be activated and switched to red in the event of an aircraft emergency, the end of one of the runways being very close to the railway line.

Bottom left. A particularly striking livery was this one promoting Great Scenic Railways of Devon and Cornwall. The black and gold scheme, with a full height picture of Truro Cathedral, was applied to six Class 153 units based at Cardiff Canton depot. As the legend on the side of the vehicle says, a major stakeholder in the promotion of these routes was the Devon and Cornwall Rail Partnership, which included Wales and West as operator, local authorities, and the University of Plymouth. Founded in 1991, it was the UK's first Community Rail Partnership, numbers of which have grown to over 70 during the years since. Working with the rail operators, they have secured more frequent services on many routes, improved station environments and facilities, and encouraged many more passengers back on to rail.

Above. One of the smaller franchises, Valley Lines, was awarded to Cardiff Railway Company, owned - like Wales and West - initially by Prism Rail and subsequently by National Express. Operations started on 13th October 1996, and covered the four routes north from Cardiff to Treherbert, Aberdare, Merthyr Tydfil and Rhymney, the southern branches to Cardiff Bay, Penarth and Barry Island, and the cross-city services between Radyr and Coryton via Cardiff Queen Street. As part of its franchise agreement, it also proposed to operate six daily longer distance services from Pontypridd, three to Manchester Piccadilly and three to Portsmouth Harbour, but these never materialised. It operated a fleet of 35 diesel multiple-units of Classes 142, 143 and 150, some of which were turned out in the company's livery of red, green and cream with the Valley Lines 'dragon' logo at the mid-point on the bodysides. The logo is seen here on unit 143 609, also carrying the name *Tom Jones*, while still in Regional Railways colours. Valley Lines was absorbed into the new Wales and Borders franchise as part of the reorganisation exercise in October 2001.

# Central Trains

Top left. The Central Trains network was based around two 'hubs', at Birmingham New Street and Nottingham. From these, long distance services radiated to destinations including Cardiff, Aberystwyth, Shrewsbury, Liverpool, Sheffield, Cambridge and Norwich. Suburban services were also operated in conjunction with West Midlands PTE, under the 'Centro' brand. The franchise was awarded to National Express on 17th February 1997 for a seven-year term, with operations commencing a couple of weeks later on 2nd March. The operator inherited a mix of diesel and electric multiple-units from Regional Railways, including the Class 323, represented here by unit 323 208, framed by the overhead catenary as it heads away from Birmingham New Street. Rolling stock deployed on West Midlands PTE services wore a modified version of the Regional Railways livery, with the dark blue band along the windows replaced by Centro green.

Bottom left. Centro liveried Class 150 unit 150 202 heads away from Barrow-upon-Soar with a Loughborough to Leicester 'Ivanhoe Line' service. Patronage does not appear particularly high; a single passenger is climbing the footbridge steps, although there may be others out of sight behind the train! In addition to the livery, stock working the PTE sponsored services also carried the yellow Centro logo alongside the doors, just visible in this view. Central Trains originally operated 38 Class 150 units, 29 in two-coach formation and 9 of three coaches. In addition to the Centro workings, they were also deployed on some longer distance services, reaching such destinations as Hereford, Matlock, Worksop and Skegness.

Below. On 4th May 1997, Centro liveried unit 323 219 calls at Barnt Green with the 15.23 Litchfield Trent Valley to Redditch service. This cross Birmingham service ran at a twenty-minute frequency from the cathedral city of Lichfield to the 1960s designated new town of Redditch, taking 90 minutes for the 33-mile journey, with 21 intermediate stops. The route was electrified in 1992/3, allowing the then newly introduced Class 323 units to be deployed on it. These three-coach units were built by Hunslet Transportation Projects from 1992 to 1995, 26 for West Midlands services, and 17 for Greater Manchester. They operate from the 25 kv AC overhead system, and are powered by Holec DMKT 52/24 traction motors rated at 196 hp, four to each of the end driving vehicles. This gives a total power output for the unit of 1,566 hp, and a maximum speed capability of 100 mph, although they have few opportunities to exploit this on their normal workings! Each driving vehicle seats 98 passengers, while the centre trailer has 88 seats plus a toilet.

Below. Central Trains' Class 158 units all started off in Regional Railways livery, with the new operator's branding replacing the 'Express' lettering on the bodysides. From 1999, the sets started to appear in this new scheme from design agency Best Impressions, seen here on an anonymous example at Nottingham. The main colours were green, dark on the lower bodyside and a strikingly bright shade on the upper. The lower bodysides carried Central's logo and branding at each end, with their phone number in the centre, replaced from around 2003 by their website address. Window surrounds were dark grey, as were the upper cab fronts, the lower half retaining the (at the time) mandatory warning yellow. A finishing touch was the blue upsweep on the upper cab sides. The operator inherited a total of 36 Class 158s at the start of the franchise in 1997, which it deployed on its longer distance services. Numbers reduced to 25 over the seven-year term, as newer stock arrived.

Top right. From 1999, Central Trains started adding new Class 170 'Turbostar' diesel units to its fleet, with 23 two-coach Class 170/5 sets and ten three-coach Class 170/6s initially. These were used to displace earlier Class 156 and 158 units from some of the longer distance workings. It subsequently acquired a further 20 units, taking its total 170 fleet to 53, but these later sets were second-hand, taken on from other operators. The initial batch were all delivered in the two-tone green livery seen on the Class 158 unit in the previous picture. In this view, passengers leave a Liverpool to Stansted Airport service, in the hands of a Class 170/5 unit, at the Rutland town of Oakham. The weather conditions have created some interesting effects here. The bright sun shining on the front of the train has completely obscured the unit number, while the sky behind looks distinctly threatening.

Bottom right. Central Trains services between Liverpool and Norwich traversed the picturesque Hope Valley route through the Peak District between Manchester and Sheffield. In this view, a three-coach Class 170/6 unit makes its way past Edale, with the village and church visible on the left. Edale marks the southern end - or starting point - of the Pennine Way National Trail, and the station is well used by walkers both 'doing' the trail, and visiting the area more generally. At the time of this picture, it generally enjoyed an hourly service, provided by First North Western Trains between Manchester and Sheffield. Most Central Trains services passed non-stop along the route, with only a limited number calling at the intermediate stations.

Way out

# North Western Trains

Above. In 1989, Regional Railways services in the North West began operating under the 'Network NorthWest' brand. Conceived along similar lines to its counterpart in the South East, this encompassed a number of marketing initiatives, including rover tickets giving unlimited travel throughout the region. A new livery, comprising a blue upper body, light grey lower body, and a red and grey bodyside stripe incorporating the 'NW' logo was applied to a small number of Class 150 units. The initiative was short-lived, however, with the sub-sector becoming the North West Regional Railways franchise in readiness for privatisation. It was awarded to Great Western Holdings - who had also acquired the InterCity Great Western franchise - on 2nd March 1997, and commenced trading as North Western Trains. The operator inherited a mixed fleet of rolling stock from BR, including some first generation Class 101 diesel units, Class 142 'Pacer' units, 'Sprinter' units of Classes 150, 153, 156 and 158, and Class 309 and 323 electric units. There were also some Mark 2 coaches for use on longer distance services along the North Wales coast, generally hauled by Class 37 locomotives hired from the freight operator English, Welsh and Scottish Railways (EWS). One of the Class 153 units, 153 359, is seen here approaching the request stop at Roman Bridge with a train from Blaenau Ffestiniog to Llandudno.

Top right. North Western's network was centred on Manchester, with routes radiating from there to Chester, Holyhead, Liverpool, Blackpool, Windermere, Sheffield and Birmingham. It also operated some relatively short-lived services to London Euston, from Rochdale and Manchester Airport. Worked by Class 322 units hired in from WAGN, these commenced in 1998 and were withdrawn by 2000. From late October 1997, around eight months after the start of the franchise, rolling stock began to appear in the company's new livery, dark blue, with large gold stars at vehicle ends connected by a gold stripe at cantrail level, and the logo and branding in a central position on the lower bodysides. The colour scheme sat particularly well on the Class 158 units, as demonstrated here on unit 158 758, pausing at Bangor with an eastbound service from Holyhead.

Bottom right. For services on the electrified network around Manchester, North Western deployed its fleet of 17 Class 323 units. Only two of these - units 323 224 and 233, the latter seen here at Manchester Piccadilly - acquired the operator's livery. The other 15 sets retained the Greater Manchester PTE colours in the early years of the franchise, and then went straight to the First Group livery following the subsequent change of ownership. Piccadilly station underwent a four-year, £100 million redevelopment programme starting in 1988, in advance of the 2002 Commonwealth Games held in the city. This involved renovation and reglazing of the trainshed roof, renewal of the main entrance and enlargement of the concourse, the provision of a new passenger lounge, and renewal of canopies and resurfacing of platforms. Some of the latter work is in progress in this view, on the platform 8 and 9 island immediately to the left of the train.

158758
north western trains

323233

Above. In July 1997, North Western placed an order with Alstom for 27 Class 175 'Coradia' diesel units, 11 in two-coach formation and the remainder as three coaches. They were built at Alstom's Washwood Heath Works in Birmingham between 1999 and 2001, with the first entering service during 2000. They were designed for long distance services, with two-plus-two seating throughout, and particular attention was paid to ride quality and soundproofing. Each vehicle is powered by a Cummins N14 diesel engine of 450 hp, giving a maximum speed capability of 100 mph. Their introduction would allow the company to withdraw some of its oldest rolling stock, including the locomotive hauled coaches. In 1998, First Group bought out the other shareholders in Great Western Holdings, and rebranded the franchise as First North Western. A new livery was designed, based on First's house colours of blue, white and pink, and this was gradually applied to the operator's rolling stock over the following years. The new Class 175s were delivered in these colours, as seen here on unit 175 005, leading an unidentified sister unit out of Conwy tunnel on the North Wales coast line.

Bottom left. . In their early days, the 175s were beset by reliability problems, centred around the bogies and brakes. Availability fell to 50 per cent and below on occasions, and the company had to resort to various measures to cover this, including hiring in locomotive hauled sets from charter operators. However, remedial action resolved most of the issues, and reliability subsequently improved. In another North Wales coast location, one of the two-coach units, 175 003, is seen here heading east between Penmaenmawr and Conwy in late afternoon sunshine. The tunnel mouth seen above the railway carries the eastbound A55 North Wales Expressway through the Penmaen-bach headland; the westbound carriageway is carried through a separate tunnel a little way to the south. The First North Western franchise came to an end in 2003/4, with services split between Trans Pennine Express, Wales and Borders, and Northern Rail.

Above. Merseyrail is the term used for the electrified suburban lines serving Liverpool and the Wirral Peninsular. The current network in the central area dates from the 1970s when new tunnels were built, one creating a circular link from the Wirral lines to James Street, Moorfields, Lime Street and Central stations, and the other linking lines from north and south to create a cross-city route, the Northern Line. Electric trains had first reached Rock Ferry, south of Birkenhead on the Wirral, as far back as 1903, taking power from a live rail at 625 volts DC. The routes north to Southport and Ormskirk followed in 1906 and 1913 respectively, while those to West Kirby and New Brighton, north of Birkenhead, were electrified in 1937. Further live rail extensions had to wait some 40 years until 1977, when the section from Sandhills to Kirkby was electrified, followed by Central to Garston (now Liverpool South Parkway) in 1978 and on to Hunt's Cross in 1983, Rock Ferry to Hooton in 1985, and finally Hooton to Chester in 1993 and Ellesmere Port in 1994. In this view, Class 507 unit 507 010 rests in one of the terminal platforms at Rock Ferry, the original terminating point for electric services from the centre.

Left. The Merseyrail Electrics franchise was awarded to MTL (Merseyside Transport Ltd) on 19th January 1997. MTL had originally been part of the Merseyside Passenger Transport Executive (MPTE), but under the bus deregulation provisions in 1985, those operations were effectively privatised, creating the new company. The PTE remained in overall control through its 100 per cent shareholding until 1993, when the company was bought by its managers and staff. MTL was then bought by Arriva on 18th February 2000, and they became the new franchisee. The distinctive yellow 'M' Merseyrail logo first appeared in BR days, but was retained by MTL at privatisation and used widely on signage and rolling stock. It is seen here at James Street, the first station in the city centre east of the Mersey tunnel.

# Northern Spirit

Top left. Original electric rolling stock was provided by the Lancashire and Yorkshire Railway on the Southport and Ormskirk lines, and the Mersey Railway on the Wirral, but from 1938 new stock, of a strikingly modern design with sliding doors, began to appear. This batch was built by the London, Midland and Scottish Railway (LMS), but a further batch, still to the original LMS design, was turned out by BR in 1956. Under the TOPS scheme, these all became Classes 502 and 503. The last of these soldiered on until 1985, being gradually replaced from 1978 by new Class 507 units, and from 1982 by the similar Class 508s, both built by BREL at their Holgate Road Works in York. The latter class spent their first three years on Southern Region suburban services, before being replaced by Class 455 units in the south and finding their way to Merseyside. Class 507 unit 507 014 pauses at Birkenhead North with a service to West Kirby.

Above. The Regional Railways North East franchise operated an extensive network of services across the north of England, serving such diverse destinations as Carlisle, Morecambe, Sunderland, Cleethorpes, and many points in between. It was awarded to MTL, who also operated Merseyrail, on 2nd March 1997. In May 1998, it rebranded as Northern Spirit, adopting a livery which comprised a turquoise base colour and a full height, heavily italicised letter 'N' in lime green a little off-centre on vehicle sides. At the start of the franchise, it had an extensive fleet of Pacer and Sprinter diesel units of Classes 142, 144, 150, 155, 156 and 158, together with a small number of Class 308 and 321 electric units. Sidelit by low afternoon sun, Class 156 unit 156 475 heads south along the Settle and Carlisle line near Long Preston with a Carlisle to Leeds service on 15th January 2000.

**Below. Class 142 'Pacer' unit 142 026 waits to depart from Carlisle with a service to Middlesbrough. As mentioned earlier, the Pacer units were one of the outcomes of extensive design studies carried out by BR in the late 1970s to replace the extensive fleets of first generation diesel multiple-units, the other being the 'Sprinter' family of units. While the Sprinters were aimed at longer distance workings, the Pacers were developed as a relatively low-cost solution for lightly used rural routes together with some suburban services. Entering service from 1985, they comprised a heavily adapted Leyland bus body - widened to take advantage of the loading gauge - mounted on a four-wheel rail chassis, and powered by a Leyland TL11 200 hp diesel engine. Without bogies and secondary suspension, track forces were transmitted direct to the vehicle body, resulting in complaints of excessive noise and rough riding. Some improvements were made to the suspension to address this, and the original engines and hydraulic transmissions were replaced by Cummins and Voith units during the 1990s. Whatever one thinks of them, they undoubtedly did the job they were designed for, and the last of them were not withdrawn until 2020, after over 30 years in service. One has to feel some sympathy with passengers boarding the train here, however, and facing a journey of around three hours to Middlesbrough, feeling every rail joint on the way!**

**Top right. Following MTL's acquisition by Arriva on 18th February 2000, the franchise was rebranded again as Arriva Trains Northern. That company's livery was gradually applied to rolling stock, first appearing on Class 153 unit 153 352 in July the following year, but much remained in Northern Spirit colours with Arriva branding. This 'hybrid' scheme is seen here on a Class 156 unit, passing a section of Hadrian's Wall alongside the Tyne Valley line. This picturesque route runs for around 62 miles across northern England from Carlisle to Newcastle. In addition to the scenery, the line also retains the only two surviving North Eastern Railway 'overhead' signal boxes, with the timber-built operating and locking room mounted over the tracks on a massive steel gantry. They are at Hexham and Wylam, both grade II listed, and both still in use.**

**Bottom right. On 3rd June 2001, Class 156 unit 156 472 stands in platform 6 at Carlisle, awaiting departure time with the 13.13 service to Leeds. The Arriva branding is clearly visible towards the far end of both coaches. With a journey time of around two and a half hours, this train will traverse some of England's most spectacular scenery along the Settle and Carlisle line, reaching an altitude of 1,169 feet at Ais Gill summit. The line was under threat of closure in the 1980s, with few passenger services, most freight using the West Coast main line, and infrastructure in a poor state due to lack of investment. However, vociferous campaigning by pressure groups and local authorities eventually produced a change of heart, and consent for closure was officially withdrawn by the Transport Secretary, Paul Channon, in April 1989. Over the years since, the line has undergone a significant renaissance, with stations reopened, passenger and freight services increased, and infrastructure and signalling enhancements carried out to cater for the greater volume of traffic.**

Above. Although most rail services in Scotland were operated by Regional Railways under the sectorised BR, they were given their own separate identity as ScotRail, which then became a single franchise during the privatisation process. It was awarded initially to National Express for a seven-year term starting in March 1997. From that date, it became responsible for most rail services within Scotland, the exceptions being those cross-border workings operated by Virgin West Coast and Cross Country, GNER, and Arriva Trains Northern. It also made some 'incursions' south itself, in the form of services to Carlisle and Newcastle, and the Caledonian Sleeper to London. The oldest stock inherited by the operator from BR was the Class 303 electric multiple-unit, of which some 40 examples remained in service out of the 91 built. Introduced in 1960 for the original Glasgow suburban electrification schemes, these three-coach units were colloquially known as 'Blue Trains', because of the smart Caledonian Blue livery in which they started life. By the late 1980s, most were wearing the black and orange colours of the Strathclyde PTE, and it is in this scheme that unit 303 013 is seen here, approaching Glasgow Central with a service from Neilston.

Top right. The blue and orange colours were replaced from around 1997 by a carmine and cream livery, seen here on Class 314 unit 314 210, arriving at Glasgow Central with a Cathcart Circle line service. Also visible on the left cab front, opposite the unit number, is the purple, red and green logo used by National Express for its ScotRail operation. By this time, Strathclyde PTE had rebranded its services simply as SPT, which later became the initials for Strathclyde Partnership for Transport. The Class 314 units belonged to the same family as the 507s and 508s built for the Merseyrail lines, but operating from 25 kv AC, rather than a DC live rail. Sixteen of these units were built in 1979/80 by BREL at their Holgate Road works in York, specifically for use on SPT's newly reopened Argyle route. From 1999, shortly after privatisation, they were deployed on Cathcart Circle services to allow withdrawal of the ageing Class 303 units.

Bottom right. The Far North. Class 158 unit 158 734 stands in the train shed at Wick station, resplendent in ScotRail colours. This is the terminus of the 161-mile line from Inverness, which also serves the town of Thurso via a short branch from Georgemas Junction, 14 miles west of Wick. This route, built in stages between 1862 and 1874, traverses some of the most remote terrain in the UK, including the 'Flow Country', the largest expanse of blanket bog in Europe. It is largely single track, although there are ten passing loops along its length where trains can pass each other. There are no lineside signals, traffic being controlled by the Radio Electronic Token Block (RETB) system from Inverness signalling centre. Successive improvements to the A9 trunk road over the years, including new bridges across the Moray, Cromarty and Dornoch Firths, have left the railway at a distinct disadvantage distance wise, the journey by road from Inverness to Wick covering just 103 miles. The typical train service is four trains per day to and from Inverness. Until 2000, services divided at Georgemas Junction with separate portions serving Thurso and Wick. However, this practice ceased with the introduction of Class 158 units, and since then all services have run via Thurso, involving a reversal at Georgemas Junction in both directions.

314 210

Welcome to Wick
158734
ScotRail

170 410

W

Top left. ScotRail acquired its first Class 170 units between 1999 and 2001, with the introduction of 24 three-coach sets numbered 170 401 to 424. These gradually replaced Class 158 units on 'Express' services between Edinburgh and Glasgow, and from Inverness and Aberdeen to Edinburgh and Glasgow. They had small first class compartments, seating nine passengers, at the outer end of each driving vehicle. A further ten similar units, 170 425 to 434, entered service by 2004, by which time the franchise had passed from National Express to First Group. In total, the operator's 170 fleet reached 59 units by 2005, most 'new-build', but with four taken on second-hand from Hull Trains. Most of these additional sets were standard class only, and were deployed on SPT services around Glasgow together with some Edinburgh commuter workings. In this view at Aberdeen, unit 170 410 rests after arrival with a service from the south. Small 'Business Class' stickers - Scotrail's branding for first class - can be seen at the tops of the first two windows behind the driver's cab.

Bottom left. Class 170 unit 170 409 approaches the up home signal at Usan with a service from Aberdeen to Edinburgh or Glasgow. Most of this route is double track, but there is a section of single track across the Rossie and South Esk viaducts between here and Montrose. This was controlled from the boxes at Usan and Montrose South until February 2010, when a small rationalisation scheme saw these two boxes closed and control centralised at Montrose North. In the background can be seen the spire of Montrose Old and St Andrews Church, which rises to 200 feet. This was built in 1834, replacing an older one on the original church building which dates from 1791. The picture is undated, but the plentiful flowering heather on the embankments places it firmly during the months from July to September.

Below. In order to complete the withdrawal of the Class 303 units on SPT services, Scotrail ordered 40 Class 334 'Coradia' units from Alstom. These three-coach sets were delivered from the company's Washwood Heath Works between 1999 and 2002. They are all standard class, the formation comprising a pair of Driving Motor Standard Open (DMSO) vehicles flanking a Pantograph Trailer Standard Open (PTSO). Total seating capacity is 183, with a two-plus-two layout in each of the driving coaches, and two-plus-three in the PTSO, which also contains an accessible toilet. Each of the driving coaches is powered by a pair of Alstom ONIX traction motors rated at 362 hp, giving the unit a total power output of 1,448 hp and a maximum speed of 100 mph. The units were plagued by early technical problems, delaying service entry until 2001, when they were deployed on Glasgow outer suburban services. They undertook some test runs to North Berwick - although they never operated there in passenger service - and it is on one of these trips that unit 334 013 was captured by the camera on the branch, looking smart in SPT's carmine and cream colours.

Above. Another Class 334 unit, 334 010, sets off from Glasgow Central with an unidentified sister unit in tow. The driving vehicles carry the SPT logo on the lower bodysides at the outer ends, while the centre trailer carries the full wording, 'Strathclyde Passenger Transport', between the doors. Glasgow Central is the larger of the city's two terminal stations, the other being Queen Street. Central has fifteen terminal platforms in the main or 'high level' section, and two more through ones in the underground 'low level' section. As well as serving the extensive suburban network around the city, the station is also the terminating point for many services from south of the border, including those from London Euston. The second ScotRail franchisee, First Group, ran services until 2015, when they handed over to Netherlands based Abellio. At the end of their seven-year term, the operation was taken back into public ownership, and is now run by ScotRail Trains Ltd, a company owned - through a holding company - by the Scottish Government.

Top right. The 1980s sectorisation programme naturally covered freight - as well as passenger - operations, with the creation of dedicated Railfreight and Parcels sectors in 1982. Railfreight locomotives were initially painted in a dark grey livery with a large BR logo on the bodysides, in one of the first breaks from the ubiquitous 'Rail Blue' scheme. This was later embellished by the addition of a red stripe, applied to the solebar or lower bodyside, depending on locomotive design. In 1987, Railfreight was divided up into six sub-sectors, each covering specific types of traffic. These sub-sectors were Coal, Construction, Distribution, General, Metals and Petroleum. At the same time, a new two-tone grey livery was introduced - lighter on the lower and darker on the upper bodysides - together with distinctive, colourful logos for the sub-sectors, each an abstract representation of the goods handled. In this view, Class 56 locomotive 56 092 carries the Coal sub-sector logo while hauling an appropriate load. The diamond-shaped plaque under the driver's side window denotes the locomotive's home depot, in this case Toton, in Nottinghamshire. The BR double arrow logo was still carried, under the cab side window at the opposite end, barely visible in this picture.

Bottom right. The spectacular Welland Viaduct, on the line between Oakham and Corby, is the setting for this view of a pair of Class 47 locomotives hauling a southbound car train. Both locomotives are carrying the Railfreight Distribution logo of two red diamonds on a yellow background, inside a rectangular red and grey border, although you cannot see this clearly in the picture. This structure, often known by the alternative name of Harringworth, after the village which sits in the valley below it, is the longest masonry viaduct in the UK. It has 82 arches, each with a 40 foot span, and a maximum height of 60 feet, giving a total length of 1,275 yards. Construction started in March 1876, and the structure was completed a little over two years later, in July 1878. Repairs carried out over the years to maintain the structural integrity of the viaduct have used a variety of brick types, giving rise to the patchwork appearance of different colours seen here. The line across the viaduct is used predominantly for freight traffic, although a handful of scheduled passenger services operate, and it is also used as a diversionary route when the Midland Main Line is closed.

# BR Railfreight sector

60098
098

048

Top left. Brand new Class 60 locomotive 60 098, *Charles Francis Brush*, stands outside its builder - Brush Traction's - works in Loughborough. It is named after the American inventor and engineer, many of whose inventions were used by the Anglo-American Brush Electric Light Corporation, a predecessor company of Brush Traction. The locomotive wears the yellow and blue 'building blocks' logo of the Construction sub-sector, whose traffic included aggregates, lime and gypsum, as well as materials used in the construction of the Channel Tunnel. The Class 60 was procured by the Railfreight sector as a Type 5 (3,000 hp plus) Co-Co locomotive which would hopefully prove more reliable than earlier types, such as Class 56. The specification called for a challenging average availability figure of 95%, and Brush Traction won the tender to build the 100-strong fleet against competition from Metro-Cammell and General Electric. The locomotive is powered by a Mirrlees Blackstone eight cylinder 8MB275T diesel engine rated at 3,100 hp, driving a Brush BA1006A alternator. This, in turn, provides power to six Brush TM2161A traction motors, one per axle. These produce a maximum tractive effort of 106,500 lbf, and a top speed of 60 mph. The locomotives were built from 1989, but the time required for rectification of numerous teething troubles meant that the first examples did not enter service until late 1990, with the last being delivered in 1993.

Bottom left. Class 60 in action. Locomotive 60 048, *Saddleback*, leans to the curve as it heads a southbound train of assorted aggregate hopper wagons along the Midland Main Line near Market Harborough. Many members of the class were named after areas of high ground in the UK, this particular example taking that of a hill in the Lake District. Following the privatisation process, all 100 of these locomotives were to end up in the hands of English, Welsh and Scottish Railway (EWS), one of the major players in the UK rail freight market. The view seen here has changed significantly since this picture was taken. Catenary masts are now in place as part of the Midland Main Line upgrade project, extending electrification from Bedford - its original 1983 limit - to Sheffield. A new footbridge has also been erected just forward of the locomotive's position, replacing a footpath crossing.

Above. Another view at Brush Traction's Loughborough Works, this time showing a Class 92 electric locomotive under construction. These locomotives were ordered for Channel Tunnel operations, specifically freight services between the Railfreight Distribution Centre at Wembley and Frethun, near Calais, together with the planned (and ultimately abortive) 'Nightstar' sleeper services from Glasgow and Edinburgh as far as Calais, where French traction would have taken over. Of the 46 built between 1993 and 1996, 30 were allocated to Railfreight Distribution, seven to European Passenger Services (later Eurostar) and nine to the French railway operator SNCF. They are dual voltage, capable of working from 25 kv AC overhead and 750 volts DC third rail supplies. Total weight is 126 tonnes, and they are rated at 6,760 hp on AC, and 5,360 hp on DC, power being provided by six Brush-manufactured asynchronous traction motors, one to each axle. They can handle a 1,600 tonne train at a maximum speed of 87 mph, although in emergency conditions, they are able to haul or propel a load of up to 2,200 tonnes out of the Channel Tunnel. Externally, the 92s bear a close resemblance to the same manufacturer's Class 60 diesel locomotives, with which they share a similar body shell. The plaque under the cab side window on the left carries the eagle logo of Crewe International depot, where these locomotives were based.

# Foster Yeoman

Above. Foster Yeoman was one of Europe's largest aggregate and roadstone suppliers, operating extensive quarries in Somerset and Scotland. It had traditionally moved most of its material by rail, acquiring its own fleet of wagons as early as 1923. This activity continued into the BR era, but the company became dissatisfied with the poor reliability of the motive power rostered to its trains, particularly the Class 56 locomotives. In 1984, in agreement with BR, the company ordered four Class 59 diesel locomotives from General Motors in Illinois, USA, adding a fifth in 1988. While remaining in Foster Yeoman's ownership, the locomotives were operated by BR crews when on the main line. Powered by a 2-stroke, V16 engine with electric transmission, the locomotives have a total output of 3,300 hp, and weigh 121 tonnes. On test, they have proved themselves capable of hauling trains weighing in excess of 4,500 tonnes. Their success in service prompted one of Foster Yeoman's rivals, Amalgamated Roadstone Corporation (ARC), to purchase four 59s of its own in 1987, and in 1993 the two companies combined their fleets to form a joint operation, Mendip Rail. A further six Class 59s were acquired by National Power from 1994, primarily for coal traffic. Foster Yeoman's acquisition and operation of its own locomotives on the national network was seen by many as a precursor to the privatisation of the whole BR network in the following decade. Here, locomotive 59 004, *Yeoman Challenger*, heads a train of empty hoppers west along the Berks and Hants line, with the Kennet and Avon Canal in the foreground.

Top right. A further reorganisation in 1988 saw the sub-sectors split between two divisions; Trainload Freight, which covered coal, petroleum, construction and metals, and Railfreight Distribution, which took in everything else. Following privatisation in 1994, Trainload Freight was sold off as three separate companies, each covering a different geographical area. The companies were Loadhaul (North East England, South and East Yorkshire), Mainline Freight (South East England, East Anglia, and the East Midlands) and Transrail Freight (West of England, Scotland and Wales). Although set up as separate legal entities with their own liveries and logos, all three were acquired in early 1996 by North and South Railways, which would go on to become EWS. Loadhaul inherited a fleet of 194 locomotives, primarily of classes 37, 56 and 60, although with a significant number of Class 08 and 09 shunters, and a Class 122 single railcar used for route learning. Main commodities handled were coal to various power stations and Scunthorpe steelworks, together with crude oil and petroleum products. On 26th April 1996, Loadhaul liveried locomotive 37 884 passes Stenson Junction, between Burton-upon-Trent and Derby, with a tanker train. (Arthur Turner).

Bottom right. Mainline Freight adopted a livery of 'Aircraft Blue', with a logo representing superimposed wheels in silver on the bodyside, and the company name below. This scheme is seen on locomotive 37 167 at Didcot on 30th October 1995, with the cooling towers of the power station behind. Note that it still carries the former Railfreight sector depot plaque beneath the cab side window, in this case Stewarts Lane, represented by Battersea Power station. The company operated a fleet of some 317 locomotives, of classes 08, 09, 31, 33, 37, 47, 58, 60 and 73, based at Toton and Stewarts Lane. Main freight flows included coal, petroleum, construction materials, and some infrastructure workings. (Arthur Turner).

## Loadhaul

## Mainline Freight

Below. Class 60 locomotive 60 084, *Cross Fell*, heads west through Newport with an iron ore train on 20th September 1995. Arguably, Transrail had the least imaginative livery of the three former Trainload Freight companies, retaining the Railfreight sector grey and simply adding their own logo and branding. Locomotives in the company's fleet included classes 08, 31, 37, 47, 56, 60. Transrail's operating area was larger than those of the other two companies - South West and North West England and the West Midlands, together with the whole of Wales and Scotland - but ironically it carried the least amount of traffic. It did, however, start up a long-distance wagonload service under the 'Enterprise' name, the first such operation since the demise of BR's 'Speedlink' service in 1991. It was only moderately successful - probably because most potential customers had made alternative arrangements once Speedlink ceased - but EWS continued it for some years after they acquired Transrail. (Arthur Turner).

Top right. The Parcels sector had been rebranded as Rail Express Systems (RES) in 1991. This was accompanied by a major restructuring of the way the sector's services - which were almost entirely supplied to one single customer, Royal Mail - were operated. This saw a move away from mail services being provided at numerous stations across the country in favour of a network based on a few major hubs, at Wembley, Tonbridge, Bristol Parkway, Stafford, Doncaster, Warrington, Gateshead and Motherwell. There was also some rationalisation of what had been a very diverse rolling stock fleet, and the introduction of some new equipment, specifically 16 Class 325 dual voltage electric multiple-units, for the carriage of mail. RES was sold in 1995 to North and South Railways, who subsequently purchased the three Trainload Freight companies as described earlier. However, traffic continued to decline for various reasons, and Royal Mail ceased all transportation of mail by rail in 2003. Much of the operator's rolling stock was painted in the sector livery of red, with a grey upper band, and light blue and grey flashes supposedly representing eagle's wings. Class 47 locomotive 47 536 wears these colours as it heads a parcels train past Dr Day's Junction, Bristol, on 31st August 1995. (Arthur Turner).

Bottom right. BR's Railfreight Distribution sub-sector was responsible for wagonload and intermodal (container) services, which had previously operated under the Speedlink and Freightliner brands, together with international traffic. At privatisation its functions were split, with the container traffic going to a new company, Freightliner (1995) Ltd, which was bought by a management buyout team backed by 3i and Electra Private Equity on 25th May 1996. The remaining operations stayed with Railfreight Distribution until 1997, when it was purchased by EWS. Thus, of the six freight operators sold off under the privatisation process, five ended up as part of EWS, leaving only Freightliner under separate ownership. This view shows an interesting cavalcade passing Clay Cross on 29th April 1996. Bringing up the rear are locomotives 90 133 and 86 604, both in Railfreight Distribution colours. Ahead of them are 87 101, *Stephenson* - the unique thyristor controlled member of the class - in BR blue, and then 92 041. The whole consist is being hauled by RES liveried 47 786, *Roy Castle OBE*. (Arthur Turner).

# Rail Express Systems

# Railfreight Distribution

Freightliner
Freightliner
66504
Freightliner

EWS
E W & S
37057
Viking

# Railtrack

Top left. As mentioned previously, Railfreight Distribution's intermodal business was split off into the new Freightliner company in 1995. The operation - and the name - have quite a heritage. Transporting goods by rail in standard shipping containers was envisaged in the 1963 Beeching report, and the first revenue earning service operated in 1965, with the 'Freightliner' name being adopted soon afterwards. With profits rising, Freightliner Ltd became a standalone legal entity in 1968, albeit still wholly owned by the government. However, significant infrastructure investment was required to allow the use of larger containers, and this led to the operation being returned to BR control in 1978, where it remained, being absorbed into Railfreight Distribution during the sectorisation process. In common with other privatised freight operators, Freightliner inherited a fleet of ageing BR locomotives, of classes 47 (diesel), 86 and 90 (electric). It soon set about replacing these, with orders for Class 57s (rebuilt Class 47s) and the American and Canadian built Class 66 - a development of the earlier General Motors Class 59 - which it bought in significant numbers. One of the latter, 66 504, is seen heading south on the West Coast main line at Stableford, north of Stafford, with a train of car carriers.

Bottom left. North and South Railways was one of the companies formed to bid for rail freight operators during the privatisation process. It was owned by a consortium whose lead member was Wisconsin Central, a railway operating subsidiary of Canadian National. As explained previously, it purchased Rail Express Systems in 1995, and the three Trainload Freight operators in 1996. In April 1996, North and South was rebranded as English, Welsh and Scottish Railways (EWS) and the holding company name was subsequently changed to match this. In 1997, it acquired the remaining operations of Railfreight Distribution after Freightliner had been stripped out, and became the biggest player in the UK railfreight market, with 90% of the business, 900 locomotives, and 7,000 employees. Locomotives were decked out in a maroon and gold livery, based very much on that of the company's transatlantic parent. This view, taken on 1st June 1996 at Swindon, shows Class 37 locomotive 37 057, *Viking*, wearing an early version; the ampersand between the second and third letters was later dropped, with the branding becoming simply 'EWS'. The company was purchased by the German Rail operator Deutsche Bahn in 2007, and the EWS identity remained for a couple more years before giving way to the name of the new owner's freight and logistics division, DB Schenker, in 2009.

Above. BR's infrastructure assets - track, signalling, tunnels, bridges, stations, etc - were transferred to a new company, Railtrack, on 1st April 1994. Early plans were that this should remain publicly-owned, but there was strong lobbying for its transfer to the private sector both from within the company and the Conservative government, and it was floated on the London Stock Exchange during May 1996. Its primary income streams were track access charges to train operators, together with leasing of stations and depots, and in its early days it performed well financially, generating annual profits and a significant increase in share value. Among the assets transferred from BR was a quantity of specialist and 'departmental' rolling stock, including the ultrasonic test train seen here, fitted with equipment for examining the condition of rails and specifically for detecting cracks at an early stage. The two driving coaches were former Class 101 DMU vehicles, 51433 and 53167, renumbered to 977391 and 977392 respectively. The centre vehicle started life as a Driving Motor Standard Open coach (DMSO) from Southern Region 4-REP unit 3015, originally numbered 62483, and now identified as 999602. In addition to the installation of the testing equipment, all three vehicles had been heavily modified and bore little resemblance to their original state.

Above. Autumn and winter seasons present the railway with a number of problems, including leaf fall, which can seriously reduce adhesion, and ice build-up on conductor rails in areas where that mode of electrification is used. The former Southern Region had a fleet of departmental electric multiple-units, converted from withdrawn passenger stock, to deal with these. This view shows unit 930 006, formerly the two driving coaches of a 4-SUB unit, now in use as a combined Sandite and de-icing unit. 'Sandite' is a mixture of fine sand and an adhesive gel, which the unit spreads on to the running rails during the leaf fall season to improve adhesion. It was adopted following experiments with other methods - a water cannon which was used to 'blast' the leaf mulch off the railheads, and plasma torches which burnt it off. The unit's other function, later in the season, is the spreading of de-icing fluid on to conductor rails, supplemented by mechanical scrapers to remove more stubborn deposits. The unit is seen wearing Railtrack's early brown and white livery. This later gave way to a dark blue and lime green scheme, while the units themselves were subsequently replaced by newly built 'Multi-Purpose Vehicles'.

Left. Unfortunately, Railtrack's early successes were to be short-lived. Outsourcing of many of its core activities to other contractors together with loss of internal engineering expertise, as a cost-cutting measure, was to lead to a number of problems. Matters came to a head with a serious derailment at Hatfield in October 2000, which resulted in four fatalities, and subsequently laid bare the extent of track maintenance failures across the network. Extensive speed restrictions were imposed countrywide, causing widespread delays and disruption for over a year while numerous similar faults were identified and repaired. The financial and reputational damage resulting from this was to lead, ultimately, to the company's demise. Amid much political and legal controversy, it was placed into administration in 2001 and liquidation the following year, when its activities were transferred to Network Rail, a not-for-profit company effectively owned by the Department for Transport. In an action shot which does show Railtrack focusing on its core activities, ballast is discharged from an MRA side-tipping wagon in preparation for the laying of new track.